WHY IS URANUS UPSIDE DOWN?

AND other questions about the Universe

FRED WATSON

WHY IS URANUS UPSIDE DOWN?

First published by Allen & Unwin in 2007

This edition published in 2008 by Summersdale Publishers Ltd

Summersdale Publishers Ltd
46 West Street
Chichester
West Sussex
PO19 1RP
UK

www.summersdale.com

Printed and bound in Great Britain

ISBN: 978-1-84024-687-2

CONTENTS

ABOUT THE AUTHOR..5

Chapter 1 RADIO ASTRONOMY..7

Chapter 2 STARGAZING..14

Chapter 3 RUNNING LIKE CLOCKWORK..43

Chapter 4 OUT OF THIN AIR..68

Chapter 5 WOULD-BE SPACEFARERS..89

Chapter 6 GREEN CHEESE NO LONGER..118

Chapter 7 MORE THAN JUST EIGHT PLANETS..144

Chapter 8 STARSTRUCK..178

Chapter 9 ACROSS THE UNIVERSE..207

Chapter 10 INDUSTRIAL-STRENGTH ASTRONOMY..225

Chapter 11 COSMIC LOOSE ENDS..257

FURTHER READING..270

ACKNOWLEDGEMENTS..273

INDEX..276

For Helen, Anna, James and Will

ABOUT THE AUTHOR

Fred Watson says he has spent so many years working in large telescope domes that he has started to look like one. He worked at both of Britain's Royal Observatories before becoming Astronomer in Charge of the Anglo-Australian Observatory in New South Wales in 1995. His main scientific interest is gathering information on very large numbers of stars and galaxies, and to this end he helped to pioneer the use of fibre optics in astronomy in the 1980s. Fred is also an adjunct professor at the Queensland University of Technology, the University of Southern Queensland, and James Cook University.

Fred is well-known in Australia for his astronomy slots at the Australian Broadcasting Corporation. In 2003, he received the David Allen Prize of the Astronomical Society of Australia for communicating astronomy to the public, and in 2006 was the winner of the Australian Government Eureka Prize for Promoting Understanding of Science.

Fred has an asteroid named after him (5691 Fredwatson), but says that if it hits the Earth it won't be his fault.

CHAPTER 1

RADIO ASTRONOMY

THE NEWS FROM THE REST OF THE UNIVERSE

Hands up if you remember 1978... If you don't, good on you. You are the future, and I hope this book might inspire you to do great things.

If you do, join the club. You're one of us. In fact, I remember 1978 as if it was yesterday. In June of that year, I made my first visit to Australia from my home in Scotland. The weather was, I recall, much the same in both places – despite the fact that they were at opposite ends of the seasonal cycle. I arrived in Australia as a young astronomer from the Royal Observatory, Edinburgh, eager to use the giant Anglo-Australian Telescope at Coonabarabran in rural New South Wales. It was then very new, and one of the largest telescopes in the world. Quite a daunting experience for a lad who'd grown up among the mills of industrial Yorkshire – or t'mills, as we really did used to call them. (There was always trubble at t'mill.)

The reason for my journey was to measure the speeds of as many stars as I could – which, oddly, I'm still doing – but it didn't work out. In fact, on that first visit, I failed miserably. The four nights I'd been allocated on the telescope were all washed out by bad weather, so I returned to Scotland empty-handed. Such meteorological calamities are the luck of the draw in astronomy, and all I could do was to reapply for telescope time the following year, hoping for better weather. The local farmers, however, were very pleased.

Looking back on it now, it's quite clear that while my return to Scotland might have lacked astronomical data, it was spiced with something else that eventually turned out to be far more important. That was inspiration – and I took it home in bucket-loads. On the staff of the Anglo-Australian Observatory at that time were three young, British-born scientists who, I'd discovered, were bursting with ideas and enthusiasm – three friends who, among other things, were collaborating together on an exciting new book. I met them all during my visit, and I was stoked. Their names were David Allen, David Malin and Paul Murdin.

Paul was actually working with me, helping me to master the various idiosyncrasies of the telescope. His main research, when he wasn't assisting klutzy visitors, was into the charred remnants of exploding stars, but he seemed interested in most other things, too. As, indeed, were the two Davids.

At that time, David Allen was making a name for himself as a pioneer in infrared astronomy, a new science that involved peering into the darkness beyond the red end of the rainbow spectrum of visible light, and seeing what could be discovered there. That early promise quickly transformed itself into a technique that today is in the forefront of cosmic exploration, propelled by modern technology. And David Malin – a former research chemist – was on the brink of creating a single-handed revolution in astronomical photography. Within a few years, David's stunning colour images of the stars would be world-famous, gracing everything from vinyl record sleeves to postage stamps.

What really impressed me about these guys, though, was the ease with which they could communicate their interest and enthusiasm to anyone – irrespective of whether they were professors of astronomy or cab drivers. All three of them had a remarkable way of explaining things, and it came from a deep knowledge and love of their subject, combined with a commonsense appreciation of what would interest others. Above all, they liked people. They were all born science communicators, and they made a deep impression on me. I yearned to be like them.

A few years later, back in Scotland, I had a chance to try my hand at this science communication lark. For my sins, as an adjunct to my research I had become manager of the Royal Observatory's Visitor Centre, so I occasionally had to field enquiries from the media. One day, there was a press release about a newly discovered cosmic recordbreaker. It was a quasar – a delinquent young galaxy – recently observed with the Anglo-Australian Telescope, and it was the most distant object known to humankind. I was invited by BBC Radio Scotland to comment on this remarkable discovery in a recorded interview for the 'Breakfast Show', and duly turned up at their Queen Street studio the evening before it was scheduled to go to air.

The interview went well, and next morning I listened attentively for my dulcet tones over the airwaves. But I listened in vain. The date was 2 April 1982, and the entire output of the British media was devoted to that morning's invasion of the Falkland Islands by Argentinean forces. The Cabinet was in emergency session, the Prime Minister was making statements, the military options were being explored... and so on. Everything else was wiped. While that morning's events almost certainly set Margaret Thatcher on the road to recovery after her honeymoon with the nation had soured, they made for a very inauspicious start to Fred Watson's broadcasting career.

Meanwhile, back in Australia, Paul and the two Davids were going from strength to strength. All of them had written popular-level astronomy books, and all were in demand as commentators on the astronomy scene. Paul moved back to the United Kingdom, eventually rising to a senior management level in Britain's astronomy programme. David Malin, firmly fixed in Australia, was producing ever more vibrant photographs of the sky using the Anglo-Australian Observatory's two telescopes. These were not only images of great beauty, but also had enormous scientific value in the detail they revealed. And David Allen's work on infrared astronomy reached new levels of sophistication with the development of innovative techniques and groundbreaking equipment – which he himself had devised.

David Allen, in particular, had extraordinarily broad scientific interests. His research in astronomy encompassed everything from nearby objects such as the Moon and Venus to the most distant quasars. As he rose to a senior level within the Anglo-Australian Observatory, his fine communication skills also gained him an increasing presence in the Australian media. His voice became a familiar feature on the airwaves of the Australian Broadcasting Corporation – the ABC. Then, disaster struck. In 1993, he was diagnosed with a vigorously growing brain tumour. The following year he died, at the age of 47. It was a tragic loss, not just for David's family, but also for science and, in particular, for the popularisation of science in Australia.

A decade and a half after the Falklands War – and three years after David's death – I got a second chance at radio broadcasting. By then, I was Astronomer-in-Charge of the Anglo-Australian Observatory, based in Coonabarabran. Out of the blue, I received an invitation to do a weekly series of early morning over-the-phone interviews about astronomy with Philip Clark, then the ABC's 'Breakfast Show' presenter in Sydney. They were ten-minute segments, eventually settling into a format of discussing two or three items of astronomy and space news, and fielding the occasional listener question. Those early morning slots are still running, and now I talk with Adam Spencer, the self-styled 'Sleek Geek', who successfully made the transition from mathematics to cool radio. In between Philip and Adam, there was a memorable four-and-a-half-year stint with Angela Catterns, who took the ratings of the 'Breakfast Show' to new heights.

Angela clearly enjoyed the segments – especially when I started talking about things happening billions of light years away. And it was Angela who dreamed up diversions such as 'Name Fred's Rooster' – a competition that arose because I used to do the show outside under the stars, not far from the chicken run. Frequently, our early-morning discourses on astronomy were interrupted by loud crowing. That competition netted something like a hundred listener entries, and gave the nameless rooster a name.

It was also during Angela's watch on the 'Breakfast Show' that a spectacular overhead transit of the International Space Station occurred just as the show went to air. I watched the brilliant speck of light from my home in Coonabarabran as it moved rapidly towards Sydney, and then explained to the listeners where to look for it. We were inundated with enthusiastic calls from people who had spotted it, including one person who told us that half a bus-load of headphone-wearing commuters had simultaneously craned their necks to look out of the window.

The Sydney 'Breakfast Show' segment spawned several other regular astronomy slots with the national broadcaster (and a couple of other stations), in which I spoke with a galaxy of talented presenters. One of these slots – a long-running monthly segment on ABC Central West and Western Plains – became a forum for conversations with listeners about virtually anything to do with the sky. This was a particularly rewarding experience, as these country folk were not only proud of their big, unpolluted skies but avid observers, keen to understand the things they'd seen both by day and by night.

Being asked all those questions, whether by city or country listeners, has been a great pleasure. When I haven't known the answer, I've said so, and checked out the details for the next show. To be honest, there's no special talent in knowing a lot about the Universe, since those of us who are professional astronomers are immersed in the details from an early stage.

It's my scribbled records of a decade of listener questions – together with a number of emails – that form the backbone of this book. Each of the 150 or so questions presented here has been asked by a lively enquiring mind, and I've tried to answer in a similar vein to the radio shows. Going into print does allow more detail, of course. But it's the questions themselves that are special, because they address the issues that people actually want to know about – rather than what we scientists *think* they want to know about.

That puts a rather unusual slant on the subject matter of the book. For example, there is far more about the Earth's atmosphere, humankind's exploits in space and the visibility of the Moon than you'd find in a standard introductory astronomy book. On the other hand, detailed travelogues on the individual planets are thin on the ground – no doubt because people are already well informed from other sources. Over the years I've also had a handful of questions that were more appropriate to a first-year undergraduate course than a light-hearted radio show. I've had no hesitation in dumping them – after looking up the answers to make sure I'd got them right on air...

There is something quite unique about being able to engage with a community via the medium of radio, and I still relish it. David Malin recently made my day by speaking of the radio slots with great warmth, praising their 'beguiling gentleness and high information content'. I do hope some of those qualities have found their way onto these pages.

So please enjoy. Whether you remember 1978 or not.

CHAPTER 2

STARGAZING

ASTRONOMY, TELESCOPES AND OBSERVATORIES

Why do we do astronomy? A colleague of mine in the United Kingdom was once asked that question, with particular regard to the public funding spent on it. 'Oh,' he replied expansively, 'it's quite simple. Astronomy is the end product of civilisation.'

Although I understood the reasoning behind his answer, I thought it was pretty arrogant. And the person to whom it was directed was clearly unimpressed, regarding my colleague with some disdain before moving off to talk to someone else. You could tell that the words 'conceited' and 'prat' were right on the tip of his tongue as he walked away, and I have to admit they weren't far from the tip of mine.

The end product of civilisation. Well, perhaps it is, but I think there are many other candidates for the honour. Endeavours that are quintessentially beneficial to humankind, with overtones of excellence, perhaps. Activities that fulfil some deep-seated need in the human spirit, but don't contribute anything to the survival or advancement of the species.

Astronomy qualifies on the grounds that it tells us about our environment on the widest possible scale. It addresses the Big Questions. Where did we come from? Are we alone? What is our destiny? It opens our eyes to the staggering Universe that exists beyond the immediate grasp of our five senses. On the other hand, it stopped being useful to humankind around the end of the eighteenth century, when most of the problems associated with navigating the Earth's oceans had been solved. Astronomers still wonder when someone important is going to notice.

It seems to me that many human activities are as well qualified as astronomy to be considered the end product of civilisation. A little while ago, for example, I found myself sitting in the Sydney Opera House, listening to a symphony concert. For some reason, I began to wonder what a visitor from Alpha Centauri would make of all the familiar, time-honoured rituals. The tuning of instruments whose origins go back two or three hundred years, the conductor in his ridiculous formal attire, those embarrassed silences when no one is quite sure whether to clap and, finally, the thunderous applause at the end. Viewed with detachment, the whole thing is quite bizarre.

But it's not useless. The audience feels thoroughly good afterwards; certainly on this particular occasion they did, since the main work on the programme was a composition by a local hero with a growing international reputation. This one man, Ross Edwards, in himself represents an end product of civilisation, as he works alone in his study in Sydney, crafting music of the most sublime beauty from nothing more than his fertile imagination and the inspiration of the Australian bush.

Of course, the same could be said of most artistic endeavours. People pay to go to rock concerts, operas, art exhibitions, movies and, at the end of the day, they feel good. We lump it all together under the heading of 'culture'. Likewise, though indirectly, people pay for astronomers to explore the Universe, and I hope the things we discover make them feel good, too. It's clear that despite what my prattish – sorry, British – colleague had to say, astronomy is far from unique in qualifying as an end product of civilisation.

Astronomy does have some particular attributes that make it worth spending public money on. First, let's be clear what sort of money we are talking about. Apart from the handful of big-ticket items noted in Chapter 11 (see 'Why should governments spend money on astronomy and space research when there are so many other needy causes?'), it's not much. At present, Australian astronomy – the observatories, radio telescopes, university departments and so on – costs every person in the country about A$3 per year. Funding for British astronomy is comparable, at about £2 per person per year. And astronomy in Britain and Australia does not do too badly from the public purse, in comparison with other nations.

But what does the public get out of it? First and foremost, we are an intensely curious species, and there is a deep-seated appeal in having our curiosity about the Universe satisfied. Humans have looked in wonder at the sky since time immemorial, and the information we now have at our disposal is nothing less than stunning. The scientists who uncover that information have a responsibility to put it into the hands of the people who, ultimately, foot the bill. It's then up to the public to decide what they want to do with it. Their response ranges from a moment's interest in an astronomy snippet at the end of a TV news bulletin to the all-consuming passion of serious amateur astronomers – and includes everything in between.

It's the in-betweens who have asked most of the questions in this book, and their interest in the way the subject works is manifest in this chapter. What is this stargazing business? How do I go about discovering more for myself? Can I get a telescope of my own? What happens at observatories? Are astronomers normal people...? And so on. It's most gratifying that in all the years I've been talking to people about astronomy, I've never once had anyone complain about the value for money they get.

Satisfying our human curiosity is only part of the story, however. Perhaps a more important reason for investing in astronomy is that a fascination with the Universe attracts gifted students to science. They might not go on to a career in astronomy, but they may well make a significant contribution in another field that brings direct benefits to society – medicine, engineering, environmental science, and so on.

At the present time, educators are experiencing unprecedented difficulties in attracting kids into the sciences – particularly the physical sciences. With challenging coursework, high fees and a rather ill-defined career structure at the end, there's no wonder that so many kids regard science as, at best 'not for me' and, at worst, nerdsville. The wellbeing of our nation is at stake here, since it relies for its success on a flow of bright, innovative scientists and engineers moving into industry and research. If the wonders of the Universe can – as they often do – get young people thinking outside the box and switching on to scientific ideas, then it has more than repaid the public investment.

Another, less obvious benefit of astronomy is that it places unusual demands on technology, which often result in solutions beneficial to other areas. Many people still think professional astronomers carry out their work by looking through telescopes when, in fact, they haven't done that for well over a hundred years. Back in the nineteenth century, photographic plates were found to be much more sensitive than human eyes in picking up the faint light of stars, and now even more sensitive TV-type detectors have taken over that role. It's second nature to astronomers to push their instruments to their very limits to reveal the deepest secrets of the Universe. Today, this means creating sophisticated equipment, the development of which places demands on industry to get the best possible results. It also results in potential benefits to other areas – for example, medical imaging, robotics and software engineering.

Less quantifiable are the secondary spin-offs of astronomy. Research facilities built by astronomers can have a very positive effect on local tourism, for example, since they are often in remote and physically spectacular areas. More mundane, but no less important, is the attention now being drawn to energy-efficient and sky-friendly lighting by the International Dark-Sky Association (IDA). The IDA itself began as a result of astronomy-minded individuals recognising that light pollution not only had a detrimental effect on observing conditions but would, ultimately, ruin the night sky for everyone – and many animal species, too – as well as running up a huge bill for what is, after all, wasted energy.

At the end of the day, however, it is politicians who vote public money into fundamental sciences such as astronomy, and there can be no doubt that there is a political dimension to the subject. Any country that can afford to devote some fraction of its national wealth to astronomy – no matter how small – is demonstrating that it has more than just the resources needed to feed its people and defend itself. Becoming a significant player in this field of endeavour makes a bold statement to the rest of the world. It also encourages scientific cooperation since, in astronomy, international borders are very low.

There's one last thing that you might consider to be a tangible benefit of studying the sky. Astronomy may eventually turn out to be *really* useful. The best example of this is our growing understanding of space. This has revealed a level of risk from so-called Near-Earth Objects (NEOs), about which you can read more in Chapter 7 of this book. While I don't subscribe to the view that the Earth will more than likely be zapped by an asteroid the day after tomorrow, I do believe there is a significant long-term threat of impact. Astronomers will continue to be involved in the quest to discover all potentially hazardous NEOs. Perhaps one day, then, astronomy will be regarded not so much as an end product of civilisation – but as the saviour of civilisation itself.

SKYWATCHING: ASTRONOMY IS LOOKING UP

What's the best way to identify the constellations visible at any particular time?

Constellations are traditional groupings of stars that help us find our way around the sky (see Chapter 8). Short of having an expert standing beside you, the best way for a newcomer to identify the constellations is with the aid of a planisphere, or star-wheel. Of course, if you can have both the expert *and* the planisphere, so much the better, but the planisphere is a really neat little gadget that is very straightforward to use.

Basically, it consists of two discs of card or plastic, one rotating behind the other. The front disc has an oval window representing the sky visible at any one time, while behind it is a star map. You simply dial up the date and time around the rim of the disc, and the window shows you exactly what the sky looks like at that time. The edge of the window represents the horizon (and is usually labelled with the cardinal points of the compass), while its centre corresponds to the point directly overhead. The idea of the visible hemisphere of the sky being projected onto a plane is what gives the little device its name. With a planisphere and a flashlight, it's very easy to identify the constellations.

It's worth remembering that planispheres are specific to a particular latitude, which should match your own latitude as closely as possible. While a few degrees' error won't make much difference, a change of hemisphere certainly will. It's also important to note that planispheres don't show the Moon and planets, which move among the stars. For the best ways to find these wanderers, have a look at the next question. Finally, if daylight saving is in force, you'll need to remember to subtract an hour from the time when you're setting the planisphere.

Planispheres are available from bookshops and some specialist stores, and come in many different guises. Among the best is a double-sided version, with a second window on the back to give you a clearer view of the constellations opposite the pole (that is, in the northern sky for observers in the southern hemisphere – and vice versa).

Once you've discovered that you can identify constellations relatively easily with your planisphere, you might find it worthwhile investing in a star atlas to give you more detail. Generations of amateur astronomers have used a famous publication called *Norton's Star Atlas*, which first appeared in 1910 and is still in print twenty-odd editions later. It's conveniently laid out, and the maps are accompanied by lists of objects suitable for telescopes and binoculars. *Norton's* isn't the only star atlas, though, and fresh-looking modern maps such as Wil Tirion's *Cambridge Star Atlas* provide excellent guides to the sky. Incidentally, both planispheres and star maps are very effectively simulated by software packages, many of which are available as free downloads. My preference is for the printed versions, however, which are rather more manageable outdoors in the dark.

Finally, planispheres have been around for a very long time in the form of astrolabes, which originated in ancient Greece, but reached a high level of sophistication in the hands of Islamic astronomers a thousand years ago. These circular brass instruments were equipped with a sighting device to measure the angular height of celestial objects above the horizon, and sometimes had interchangeable plates representing different latitudes. An echo of this can be found in a gadget made for Allied airmen during the Second World War, the curiously named 'Rude Star Finder and Identifier', which was a planisphere with seven interchangeable templates for different latitudes. Back in those dark days, air navigators needed all the help they could get in finding the stars.

What's the best book to buy to tell me how to find the planets?

As noted in the previous answer, the Moon and planets (and the Sun, too) all move against the background of stars, and thus can't easily be depicted on a planisphere. This is one area where software maps of the sky come into their own, since the planets' positions are already programmed in. Packages such as *MegaStar*, *Redshift 5*, *Starry Night*, *TheSky* and *Voyager 4* effectively simulate the night sky. Likewise, web-based sources such as www.heavens-above.com provide planetary positions.

There are books, too, that show you how to find the planets, in the form of almanacs, or yearbooks. These publications contain details of sunrise and sunset times, the phases of the Moon, eclipses, and so on – as well as the positions of the planets and, often, other Solar System objects. Basically, they contain everything that isn't the same from one year to the next – which the stars themselves are.

Almanacs come in a very wide range of formats and levels of detail. As some of the information they provide is specific to a given locality, different countries tend to have their own. Australians, for example, are fortunate to have two excellent publications that nicely complement each other, the *Australian Sky Guide* (Powerhouse Publishing) and *Astronomy Australia* (Quasar Publishing), aimed more at amateur astronomers.

Many leading astronomical societies produce their own yearbooks – for example, the *Observer's Handbook* of the Royal Astronomical Society of Canada. This, too, has a fine reputation, and used to be known as 'the Bible according to Bishop' in honour of its long-time editor, Roy Bishop. It celebrates its centenary in 2008. The United Kingdom's principal offering is Sir Patrick Moore's venerable *Yearbook of Astronomy*, which first appeared in 1962 – to the great delight of a teenage Fred Watson. Besides maps, monthly notes on the planets and tables of useful data, each edition of the *Yearbook* contains a selection of articles on subjects of current interest. A rather older Fred Watson has been contributing to the *Yearbook* for more years than he cares to remember.

Finally, if you don't want to buy a book or a software package, you can find information on the planets in many of the colourful astronomy magazines that adorn most newsstands. Such publications include *Astronomy Now* in the United Kingdom, *Sky & Telescope* and *Astronomy* in the United States, *Sky & Space* and *Australian Sky & Telescope* in Australia, *Ciel & Espace* in France – and so on. You pays your money and you takes your choice...

What's the best telescope to buy?

By a fairly healthy margin, this is the question I've been asked most often by radio listeners. Unfortunately, it doesn't really have a straight answer, since there's now a very wide range of telescope types, sizes – and prices – available. However, there are some general principles to bear in mind.

When I first started taking an interest in astronomy, the question simply didn't arise unless you were *very* rich. Most budding astronomers had to make their own telescope, which often meant grinding and polishing a mirror (commonly 6 inches, or 150 mm, in diameter). Conveniently, however, there had been a world war relatively recently, so every town had its government surplus shop where you could buy lenses, prisms and other bits and pieces suitable for incorporating into your telescope. As Spike Milligan used to say, you could buy them from the surplus army store – or direct from the surplus army itself.

Today, the picture couldn't be more different. Manufacturers across the world are falling over themselves to persuade you that their telescopes are the very best you can buy, and offering fantastic deals in the process. China's recent entry into this market has undercut prices even more, and a telescope should now be within the reach of anyone aspiring to look at the sky – without them having to manufacture it themselves or pay the Earth.

The only ones to avoid if you can are those diminutive refracting telescopes (ones that have a small lens at the front, like binoculars), which turn up in supermarkets and speciality stores. They usually offer phenomenal magnification at a rock-bottom price. The fact is, however, that light-collecting power is more important than magnification and this is determined by the diameter of the lens. Moreover, the optical performance of these telescopes invariably leaves a lot to be desired, and it's almost impossible to point them steadily at the object of interest because of their flimsy mountings and spindly tripods. I often wonder how many would-be astronomers have had their interest stifled by the frustration of trying to use one of these things. If the rock-bottom price is all you have available, I'd recommend you buy binoculars instead (see the next answer).

The best instruments for astronomy are reflecting telescopes, which use a mirror rather than a lens as the main light-collecting element. Once again, it's the diameter of the mirror that determines the power of the telescope. Many American manufacturers still quote diameters in inches, so imperial units remain the parlance of choice for small telescope owners. While a 6-inch (150-mm) telescope would have been a healthy size 40 years ago, 14-inch (350-mm) telescopes are now commonplace and some serious amateur astronomers have monsters up to 36 inches (900 mm) in diameter.

Many people don't realise that a telescope's magnification can be changed by using different eyepieces, which can be bought as separate accessories. This allows the observer to get the best performance out of prevailing atmospheric conditions, or to optimise the view for the type of object being looked at. Planets, for example, usually need a higher magnification than faint galaxies.

The way telescopes are put together has also changed. Reflecting telescopes can be built with a wide range of optical layouts to get the light from the mirror to your eye, but the commonest is the so-called Newtonian form, named after the famous chap who not only hit on the idea but managed to make one in 1668. In Newton's setup, the starlight reflected from the main mirror is diverted by a small angled mirror through the side of the tube, where it can be examined with an ordinary eyepiece. This avoids your head getting inconveniently in the way of the incoming starlight.

Over the past couple of decades, Newtonian telescopes have commonly been supported by box-like structures rotating on Teflon pads in order to allow them to be pointed easily around the sky. These replace more refined and expensive mountings, allowing more of the available resources to be devoted to the vital optical components. The idea came from a passionate American amateur astronomer, John Dobson, who calls it a 'sidewalk' telescope because of his work on the sidewalks of America in giving passers-by a look at the sky. However, everyone else in the Universe calls it a Dobsonian in honour of its inventor.

Most manufacturers now produce Dobsonian telescopes in a range of sizes, and it is probably the most cost-effective option for a newcomer to the hobby. Starting reasonably small and gradually progressing up the size range is often a good idea, as the size dictates the price. All reputable makers produce telescopes of good quality, but it remains true that you get what you pay for. The best way to check out what is available and how much it will cost is to buy one of the astronomy magazines mentioned in the previous answer. They are full of ads for telescopes and suppliers, and frequently have test reports on new equipment, or best-buy comparisons. They will also point you to manufacturers' websites, which contain a wealth of information on their products.

Are binoculars any use for stargazing?

You bet they are – particularly for objects that cover a largish area of sky, and especially when said objects are viewed on moonless nights away from city glare. Prime targets include star clusters, nebulae (misty looking patches that can be anything from gas clouds to galaxies) and occasional bright comets. Binoculars come into their own when you're scanning the Milky Way, giving a real sense of the disc of our Galaxy (see Chapter 8). And, of course, they're also great for daytime use.

There are various trade-offs between magnification and lens size when you're choosing binoculars. Particularly important for astronomical viewing is the diameter of the objective lenses at the front, since they determine how much light the instrument can gather and squeeze into your eyes. A standard night glass is 7×50 (7× magnification and 50-mm diameter objectives). That's a good combination if you've got really dark skies. It's especially powerful if you're young enough for your eye pupils to expand to the full 7-millimetre diameter of the light beam emerging from each half of the binocular, once your eyes are fully dark-adapted. (Unfortunately that ability begins to decline when you reach the age of about 40.)

I think that going to a slightly higher magnification (say 10×40 or 10×50) is worthwhile, however, both for the extra detail you see and the fact that it's a little more forgiving of artificial light pollution. On the other hand, the higher magnification makes it harder to hold the glass steady without a stand. Some modern binoculars, such as those produced by Canon, incorporate an image stabilisation system that dramatically improves the steadiness of the images. In general, with binoculars as with telescopes, you get what you pay for, so it's worth going for quality. Regrettably, not many of us can afford a top-line Pentax, Nikon, Swarovski, Zeiss, Leica, etc., etc. but at least there are some more affordable options out there. Again, it's worth checking the astronomy magazines.

What's a good present for an interested child of 6/8/10/12 etc?

I hope the preceding answers give some good hints on the answer to this question. A planisphere makes a fantastic present for a kid of almost any age – an unusual gadget that allows them to engage with the sky, no matter where they live. Best of all, it doesn't need batteries. Telescopes and binoculars are much more expensive, but are often bought as presents for the whole family, nurturing an interest among the kids.

There's also a very wide range of appealing astronomy books for children, and by the time you get to the upper end of this age range, the standard fare of astronomy reading is quite appropriate. That includes works by Sir Patrick Moore, Dava Sobel, Ken Croswell, David Malin, and so on.

Something else to think about for a birthday or other special occasion is to go along to an astronomy viewing night. Many towns and cities have active astronomy clubs, most of which hold occasional evening viewing sessions that the public can attend. These give people the opportunity to look at the sky through sizeable telescopes without having to invest in any equipment. It's also a great way for newcomers to begin finding their way among the constellations in the company of people who truly know and love the night sky.

While everyone wants to encourage an interest in astronomy from an early age, I recently heard of a case where this might have been taken just a little too far. It was at one such astronomy viewing night that a friend of mine held a toddler up to the eyepiece of his telescope for a look at the planet Mars. But instinct took over, and instead of looking through the eyepiece, the youngster latched onto it with his wide-open mouth and gave it a good suck before my friend had time to realise what was happening. I can't imagine it tasting very good, but who knows what future interest in astronomy might have been triggered?

DOMES AND DISHES: TOOLS OF THE TRADE

Why do you build telescopes on mountain tops rather than in flat deserts?

Actually, some types of telescope *are* built in flat deserts, but first we have to clarify just what professional astronomers mean by a 'telescope'. Until 1932, when Karl Jansky of the Bell Telephone Laboratories discovered radio waves coming from space, there was only one kind. It collected and focused ordinary visible light, using the science of optics to retrieve information – originally by human eyes then, from the 1880s onwards, by photographic plates and films.

Such telescopes are still built and used today, albeit with sensitive TV-type detectors 20 or 30 times more effective than photographic films – and with mirrors that range up to *10 metres* in diameter. If you think that's big, instruments more than twice that size are on the horizon. But these 'conventional' telescopes have also been joined by as many other kinds of telescope as there are varieties of natural radiation traversing the Universe.

These ghostly emissions are gamma rays, X-rays, ultraviolet rays, visible light, infrared, millimetre and radio waves. You can imagine them all as waves with differing peak-to-peak separation, or wavelength. Arranged in order of increasing wavelength (as listed above), they form the so-called electromagnetic spectrum, and visible light, with its own rainbow spectrum of colours, is somewhere near the middle.

We now know that the Earth is constantly bathed in radiation covering the whole of the electromagnetic spectrum, coming from sources everywhere in the Universe. Most of it is absorbed by the atmosphere and never reaches the surface of the planet. If you want to observe X-rays or gamma rays, for example, you have to fly specialised telescopes on spacecraft. But for some categories of radio and infrared radiation – and visible light, too, of course – the observations can be made from the ground.

To distinguish them from their more exotic cousins, telescopes that use ordinary visible light are now called optical telescopes. They require clear skies and darkness to operate, so observing with them is always night work. Despite the effectiveness of the newer types of astronomy, these optical telescopes still play a vital role in the study of the Universe because of visible light's central position in the electromagnetic spectrum, and the fact that ordinary stars emit most of their energy as visible light.

Large optical telescopes are always built on mountain tops. The reason for this is that they are particularly sensitive to the effects of turbulence in the atmosphere, which spoils the sharpness of the images they can see – and the higher you go, the less turbulence there is (see 'Why do stars twinkle?' in Chapter 4). Of course, the very best place for an optical telescope is in space, where there is no atmospheric turbulence. That is why the 2.4-metre Hubble Space Telescope has been so successful. Such projects are extremely expensive, however, and are essentially one-off investments.

Back on Earth, it turns out that some places tend to suffer more atmospheric turbulence than others. During the 1960s, scientists all over the world embarked on site-testing programmes to establish where the best observing conditions could be found. They were spurred on by the new generation of wide-bodied jets that promised easy access to remote facilities. Before this, telescopes had been built wherever the astronomers happened to be – which was seldom a good spot.

A few places with the rare combination of clear skies, freedom from artificial light and low atmospheric turbulence that is the astronomer's ideal were found. Typically, these places were in middle latitudes (between 20 and 40 degrees North or South of the equator), on mountain tops higher than about 3500 metres, and near the eastern boundary of an ocean. If the mountain peak was on an offshore island and streamlined with respect to the prevailing wind, so much the better.

In the northern hemisphere, such sites were found in the south-western United States, Hawaii's Big Island and the island of La Palma in the Canaries. Continental Europe was largely ruled out because of its bright lights and indifferent weather. In the South, the peaks of northern Chile and the high Karoo of southern Africa were favoured. With no high mountains on its western seaboard, Australia could boast no world-beating sites – although the Siding Spring Observatory in north-western New South Wales is the best in the country, and still one of the world's least-polluted by artificial light.

So much for optical telescopes, but where do you build the other important type of ground-based telescope – the one that uses radio waves to study the Universe? It turns out that such instruments have quite different requirements from optical telescopes. There's no need to sit on a mountain top, for example – at least, not for mainstream radio astronomy. What is important is the site's radio-quietness – its freedom from interference by telecommunications. Thus, remote desert areas are quite good candidates for large radio telescopes. While no one could describe the pastoral setting of Australia's 64-metre Parkes radio dish as a desert, the rolling terrain is effective in shielding the telescope from human-made radio interference.

Australian radio astronomers are looking at somewhere more barren for their next big venture, however. In 2007 the Federal Government awarded just over A$50 million for the construction of an instrument called ASKAP – the Australian Square Kilometre Array Pathfinder. This is a linked array of up to 45 radio dishes that will be spread over a large area of remote inland Western Australia, providing astronomers with a unique tool for studying the early Universe and exotic spinning stars. More importantly, perhaps, it is hoped the instrument will be a forerunner of the Square Kilometre Array (SKA), a 17-nation venture that will provide astronomers with a million square metres of collecting area for their most challenging research. This 20-year project has a price tag of A$1.8 billion, and Western Australia is one of two short-listed candidate sites – the other being South Africa and its neighbours.

Is there any possibility of a major observatory being built in Antarctica?

There is, indeed. Even though night only darkens the continent for half the year, it turns out that parts of Antarctica's high plateau have very good observing conditions for optical astronomy (see previous answer). A site called 'Dome C', which is 1100 kilometres inland from Australia's Casey Station on the coast of Antarctica, has been investigated over recent years by a team from the University of New South Wales – who seem to like being freezing cold.

It turns out that once you get a few tens of metres above the ice at Dome C, the level of atmospheric turbulence is possibly the lowest on the planet. A telescope there could deliver images rivalling those of the Hubble Space Telescope in sharpness – but for a fraction of the cost of putting a telescope into space. Pilot studies into the feasibility of building a large observatory are continuing.

Can you build telescopes with liquid mirrors?

Yes, you can. But why should you want to do it? And how do they work? The heart of a large optical telescope is its mirror, and without it the instrument would be useless. To be exact, it's the front surface of the mirror that is the vital part – the rest of the material from which the mirror is made merely serves to support a microscopically thin layer of aluminium or silver that reflects the light to form an image. Unlike ordinary mirrors such as the ones in your bathroom or your car, telescope mirrors are reflective on their front surfaces rather than at the back.

The front of a telescope mirror is made in the form of a shallow dish, and its shape has to be precisely controlled during manufacture. For example, the 3.9-metre diameter mirror of the Anglo-Australian Telescope has a surface accuracy of about 0.000025 millimetres. That means that if it was expanded to be the size of the British Isles, the biggest irregularity on it would be about the height of a pencil laid on its side. Pretty smooth. As you can imagine, manufacturing such a component is quite a challenge.

As telescopes have become larger, different methods of making their mirrors have evolved. Some of today's largest telescopes use mirrors that are made not of a single piece of glass, but of smaller hexagonal segments butted together to make a continuous surface. They have been successful in sizes up to 10 metres, and this will be an important technology for the 20- to 30-metre diameter telescopes of the future, the so-called ELTs (Extremely Large Telescopes – what else?).

A number of scientists over the last 150 years or so have spotted a way of circumventing all this hard work by making a telescope mirror out of a rotating dish of liquid mercury. The reason this can work is that when you rotate a liquid, its surface takes on exactly the shape required for a telescope mirror. This little snippet of information was mostly of academic interest until 1982, when Ermanno Borra of Quebec's Laval University set about demonstrating the feasibility of building large telescopes using liquid mirrors. One obvious handicap is that such telescopes can point only directly upwards, or you lose all your mercury. But there are ways of working around this to give the telescope a small patrol area around the zenith (the point directly overhead) and, of course, as the Earth rotates different stars are brought into view.

Borra's pioneering work has led to a new observatory at the University of British Columbia in Canada. The major instrument there is the Large Zenith Telescope, which has a liquid mirror 6 metres in diameter. A film of mercury only a few tenths of a millimetre thick is supported on a rotating former that approximates to the correct shape. That minimises the amount of mercury required, reducing the mirror's weight and toxicity. Those two issues have led to the investigation of other exotic reflective materials, such as gallium-indium which is liquid above 16°C.

AIMING HIGH: ASTRONOMERS AT WORK

What's a billion?

Good to have this cleared up at the outset. When I was a lad, there were two sorts of billion. One was the British billion, which was a million million. The other was the American one, which was a thousand million. People used to say that the billion was the only thing that was bigger in Britain than America. Sadly, though, the British billion has now become all but extinct, and is today usually known by the American name of a trillion. So – in this book, as elsewhere, a billion means 1,000,000,000. In the clever mathematical notation so beloved of scientists, it's 10^9. In anyone's terms, though, it's a hell of a lot.

How are astronomical objects named?

In Chapter 7 of this book we will meet the supposedly infamous International Astronomical Union (IAU), which was responsible for demoting poor little Pluto to the status of a dwarf planet in 2006. This body doesn't spend all its time deliberating on the definition of a planet, however. All authority for naming celestial objects of any kind is vested in the IAU, whose various committees decide on the proposals for new names that are submitted to them. In that respect, we have come a long way since the early nineteenth century, when common usage among astronomers decreed that the recently discovered seventh planet should be called 'Uranus' rather than the competing suggestions of 'Herschel' (after its discoverer) and 'George' (after Britain's loopy monarch). 'George' would have been far safer.

Stars, as we will discover in Chapter 8, are traditionally named with Greek letters and the constellation name in which they are found. By the time we get to fainter stars, the Greek letters have run out and we revert to numbers called Flamsteed numbers – after the first Astronomer Royal, who initiated the practice in the early eighteenth century. Fainter stars still are usually identified by their number in a catalogue, such as the

'Henry Draper Catalogue' of 225,000 stars, completed in 1924. Star numbers in this catalogue are preceded by 'HD'. Note that commercial organisations claiming to hold naming rights for stars – which they then sell on to individuals – are being less than honest since, in fact, only the IAU has this authority.

Like stars, the misty patches in the sky known as nebulae have various catalogue numbers. The first was Messier's catalogue of 1774 (catalogue numbers prefixed with 'M'), but the most commonly used is Dreyer's 'New General Catalogue' of 1888 (prefixed with 'NGC'). Unusual objects such as variable stars are given special names referring to the constellation in which they are found, while transient objects such as supernovae (exploding stars) incorporate the year of the event in their names. Some stars are known by their number in an individual astronomer's catalogue of unusual objects – for example, the red dwarf star Gliese 581, which is now known to have at least two planets orbiting around it.

In the Solar System, only comets take the name of their discoverer, although asteroids can be named *by* their discoverer – if the IAU approves. On their discovery, asteroids and remote Kuiper Belt objects (see Chapter 7) are given a temporary identifier consisting of the discovery year, a two-letter code (the first character of which denotes the half-month in which the discovery was made) and a running number. Thus, the Kuiper Belt object discovered in December 2004 and originally nicknamed 'Buffy' (after the TV vampire-slayer), was 2004 XR 190. Such Kuiper Belt objects don't keep their nicknames, however, because the convention is to name them after deities in the indigenous creation mythology of the discoverer's home district.

This naming process can take some time, because a permanent name is not bestowed until a reliable orbit has been determined for the object in question. Kuiper Belt objects move so slowly against the background of stars that it could take a few years for this to happen.

Can you explain why astronomers talk in terms of North, South, East and West when referring to images?

What an excellent question. If you were looking at an image of Mars, for example, you'd be seeing an object that is known to be spherical like the Earth, and you could apply the same conventional compass directions to the planet. But many objects in deep space are of unknown geometry, and all we see when we view an image is a two-dimensional representation. What astronomers do in that situation is apply the directions imposed by the sky itself to the image. Thus, the direction to the North Pole of the sky is 'North' and the direction to the eastern horizon is 'East'.

What confuses many people is that if you plot these directions on a picture of the celestial object in question, they come out back to front compared with the normal compass points. That's because while they are indeed directions on a sphere, they're on the sphere of the sky (the celestial sphere), which we see from the inside. Thus they're the opposite way around to normal geographical directions.

How do you measure distances in space?

This is the most basic question in astronomy. All we see when we look at the sky is a lot of bright lights that are effectively on the inner surface of a sphere – there are no clues whatever as to their distance from us, and trying to guess them is a risky business. I once fell headlong into this trap, convincing myself that an object I could see moving silently through the dark sky of a small Australian country town was a re-entering spacecraft. It was bright, and trailed flames behind it. What else could it be?

When I volunteered this hypothesis to a gentleman standing nearby, he quickly put me right. 'No, mate. It's a garbage bag with a fire-lighter tied underneath it on a bent coat-hanger. Makes a hot air balloon, see? Someone's been launching them for a couple of weeks now, and the police are trying to catch 'em before they set the whole bloody town on fire...' I'd made the classic mistake of confusing a small, slow-moving object 50 metres above the ground with a large, fast one a hundred kilometres away. So – I'd be most grateful if you didn't let this story go any further.

It was not until Johannes Kepler worked out his laws of planetary motion, in the early seventeenth century, that some idea of distances in space began to emerge. Until then, people had known that everything was a long way off, but weren't sure how far. Kepler's work, together with the yardstick eventually provided by observations of rare transits of Venus across the Sun's disc, enabled scientists to determine the scale of the Solar System.

It wasn't until 1838 that the first distance to a star was measured, however, and the method used is explained in Chapter 8 ('How do you measure light years?'). And with that same method – and by using various other calibrations – we can work out the scale of remote objects such as galaxies by measuring the way their light is shifted towards the red end of the spectrum (see 'How do you measure look-back times?' in Chapter 9).

How can you stop light pollution getting worse?

As we saw earlier in this chapter ('Why do you build telescopes on mountain tops rather than in flat deserts?'), one of the main requirements for an optical observatory is a site that is free from artificial light. There are many other considerations, but that first hurdle was what directed the overall strategy of astronomers in the 1960s and 1970s when they were looking for the world's best places for building large telescopes. Over the years, an estimated US$3.5 billion has been invested in observatories on these sites. Given the ubiquity of urbanisation of our planet, the real challenge is to keep them dark.

In the 1980s, Roy Garstang and David Crawford in the United States did fundamental work on the spread of urban sky glow, work that has been taken further by the studies of Pierantonio Cinzano at the University of Padua in Italy. Cinzano has mapped the world's light pollution using night-time images from space (to reveal upward-pointing light sources) combined with the atmosphere's known characteristics in spreading light around by scattering effects. The process is highly refined, taking into account the curvature of the Earth, and shows that two-thirds of the world's population live in light-polluted conditions. More dramatically, about one-fifth of the population can no longer see the Milky Way.

Cinzano's studies also show that light pollution is encroaching on many of the world's major observatory sites. This leads to the conclusion that remoteness is not enough to protect an observatory, and steps have to be taken to reduce the spread of light pollution. Such steps can be legislative (and the environment of most of the world's leading observatories is so regulated) or educational – to bring home the message that light pollution is bad news not only for skywatchers, but for all of us who are concerned about wastage of energy. Light pollution also has adverse effects on the animal kingdom, including the disruption of nocturnal species and the disturbance of migration routes.

Perhaps it was a combination of these aspects that in 2002 led the Czech Republic to become the first country in the world to propose national legislation protecting the night sky. The law-makers defined light pollution as 'every form of illumination by artificial light that is dispersed outside the areas it is dedicated to, particularly if directed above the level of the horizon'. The proposal came about because of popular concern about the glare caused by poorly designed outdoor lighting. Fully shielded light fixtures – which prevent the light escaping above the horizontal plane – will be mandatory in the Czech Republic. Such energy-efficient and low-pollution light fittings are gradually becoming standard practice in other countries.

The telescopes at Siding Spring Observatory near Coonabarabran also require protection from light pollution. The Anglo-Australian Telescope, for example, is so sensitive that in moonless conditions it is capable of recording a light level equivalent to a 100-watt bulb at the distance of the Moon (384,000 kilometres). But that only works if the sky is perfectly free from artificial sky-glow. The biggest threats to this are from Coonabarabran itself (population 3,000 at a distance of 20 kilometres), the city of Dubbo (40,000 people at 100 kilometres) and Sydney (four million people at 330 kilometres).

The observatory site is protected by a legal instrument called the Orana Regional Environmental Plan No.1 (REP), which originally covered an area of radius 100 kilometres, centred on the Anglo-Australian Telescope. This New South Wales State legislation was enacted in 1990, but is now being revised. The purpose of the revision is to bring the legislation up to date (particularly with regard to the important role of the fully shielded light fixtures); to make it easier to apply; and to make it easier to understand. In addition, the area of coverage will be increased to a radius of 200 kilometres to include major population centres such as Tamworth. That city – the 'City of Light' – anticipated the REP's provisions by adopting a very effective local lighting strategy in 2002, becoming the first regional centre in Australia to do so.

Astronomers in general take a pragmatic view of this problem. They do not necessarily want to see lights switched off – but they do want to see them better-designed. For astronomers, it is a question of winning hearts and minds, and the developers, makers and sellers of outdoor lighting fixtures are their prime targets. It is no exaggeration to say that in some degree at least, these people hold the future of optical astronomy in their hands. And, perhaps even more importantly, they can help to reclaim the night sky for everyone.

How can I get a job in astronomy?

If you're at school, the message is to work hard. Particularly at mathematics and the sciences, but it's also important not to neglect English, history, geography and a foreign language. Astronomers need to be able to communicate well – and to be as clued-up about the world in which they live as they are about the Universe. Do the very best you can in your exams, and that will hopefully take you to university to study mathematics, physics, astronomy or astrobiology – depending on your aptitudes and ambitions. At the end of your degree course it's usually crunch time. Do you still want to pursue astronomy? If so, and if your degree is good enough, you will then embark on postgraduate studies – a major research project that will lead to your PhD (Doctor of Philosophy) or an equivalent degree.

Having succeeded in that, you're an astronomer – but now you have to get a job. The first post is usually a postdoctoral fellowship, after which you might expect to find a lectureship or a position in one of the national astronomy facilities. Don't expect to make a lot of money, though. In the grand scheme of things, astronomers aren't particularly well paid, but the satisfaction that comes from helping to push back the frontiers of knowledge about the Cosmos more than makes up for that.

And there are a few perks, such as the prospect of living overseas (which is almost inevitable at some point in an astronomer's career) and opportunities to travel.

What if you're already on another career path, but would still like to be involved in the science of astronomy? Don't give up hope – there are openings at observatories and universities for people with a very wide range of skills. Mechanical engineering, electronics, software engineering, public relations, technical writing – such opportunities are not commonplace, but they do exist. And it's never too late to beef up your knowledge base with an external degree course or similar mature-age qualification.

That's actually how I became a working astronomer. My university career was clouded by a less-than-spectacular performance in mathematics. To be honest, I was crap, and came out with a very poor degree as a result. But over the years I pulled myself up by my bootlaces through various jobs in astronomy, gaining first a master's degree, and eventually doing an external PhD with the University of Edinburgh. By the time I got it, I was into my forties. But it's never too late, folks.

What's the ratio of female to male astronomers?

Australian astronomers are fairly well organised when it comes to looking at the big picture. Every ten years, they carry out a Decadal Review of the state of the science, and this is then incorporated into a plan for the next decade. The body responsible for coordinating all this is the National Committee for Astronomy of the Australian Academy of Science, and their most recent thinking is presented in a comprehensive and rather beautiful document entitled *New Horizons – A Decadal Plan for Australian Astronomy 2006–2015*. It's because of the work that went into this that I can give you an exact answer to this question

– as of November 2005. While the figures refer to Australian astronomy, they are reasonably representative of the rest of the world, the United Kingdom in particular.

The bottom line is that 20 per cent of positions in astronomy are held by women, so the ratio is one to four. Perhaps the best thing that can be said about this is that it's an improvement on the way things used to be – in 1995, only 11 per cent of astronomy jobs were held by women. The better news is that the improvement is likely to continue. At the time of the 2005 Decadal Review, female postgraduate students (see the previous answer) amounted to 37 per cent of the total, an increase from 15 per cent in 1995. Since postgraduate students represent the next generation in any field of study, this can only be a good sign.

The women represented by these figures are on a completely equal footing with male astronomers. That contrasts with the situation in the late nineteenth century (and well into the twentieth), when women were used as 'computers' to carry out repetitive tasks ranging from numerical calculations to star measurements from photographic plates. They were preferred over men because of their greater care and attention to detail. One famous group of women was recruited by Edward Charles Pickering (1846–1919) at Harvard College Observatory in the United States to work on the 'Henry Draper Catalogue' (see 'How are astronomical objects named?'). They were known as 'Pickering's Harem', and included Henrietta Leavitt and Annie Cannon – both of whom eventually made fundamental discoveries in astronomy.

AND FINALLY...

What do astronomers do on cloudy nights?

Well, radio astronomers just get on with the job, since radio waves (other than short-wavelength ones) are unaffected by cloud. However, optical astronomers can't work, so the telescope has to be shut down and the dome closed. Most astronomers still have 1001 other things to be getting on with, though. Night observing usually represents only a very small fraction of their working lives – perhaps 10 per cent or so. The rest is analysing earlier observations, writing up the results for publication, applying for telescope time, applying for research grants, preparing lectures, building instruments, supporting the work of other astronomers, answering enquiries and requests, and so on. As in any other profession, astronomers are often saddled with management responsibilities as they progress in their careers. All that 'stuff' tends to spill over into cloudy nights.

A rather similar question often asked by radio listeners is 'Do astronomers do different things when the Moon is full?' The answer here is 'Yes', because there's far too much background light in the sky for faint objects to be observed successfully. What happens is either that bright objects are observed with special equipment that allows their light to be analysed in great detail, or that infrared ('redder than red') observations are carried out. Either way, the ill-effects of moonlight are much reduced. It gives a whole new meaning to the term 'moonlighting'...

CHAPTER 3

RUNNING LIKE CLOCKWORK

THE MECHANISMS OF PLANET EARTH

We inhabitants of planet Earth are the victims of a truly stunning illusion. It is that the commonplace – the everyday world we take for granted – is, well... commonplace. In the grand scheme of things, it is most definitely not. Most of the Universe consists of the cold vacuum of empty space, and those parts that aren't empty are so different from anything we have ever experienced that they can only be regarded as hostile to human life.

We have evolved as creatures of the Earth, so we are bound to feel at home here. Stuck to our planet by gravity, we breathe air that is made palatable by the greenery around us. We are sustained by the products of a long and complex food-chain. So complete is our adaptation to our environment that for thousands of years we humans have had time to devote to enterprises other than the basics of survival and procreation.

Among them is the activity you're engaged in now: asking deep and meaningful questions about our wider environment – and hopefully receiving a few answers.

Why are we so taken in by this illusion of the commonplace? Or, to put it another way, why has it taken so many millennia for us to recognise the true nature of planet Earth? The first and foremost illusion is that the Earth is essentially flat, and this comes about because of its large size compared with our own physical dimensions. At 12,756 kilometres in diameter, our world is too big to reveal its curvature without careful measurement or optical instruments. Of course, we now know that in comparison with most things in the Universe, it is tiny – the 'small blue dot' of Carl Sagan's *Cosmos*. Its diameter amounts to only 1 per cent of that of even a modest star such as the Sun.

A fellow-conspirator in this illusion is the pull of gravity. This force overpoweringly defines the direction to the centre of the Earth rather than some more significant direction in space, such as the line along which we are travelling. Thus we have a strongly topocentric ('centred on the place') view that allows us to perceive intuitively only up, down and sideways.

We also live in a world that appears to be absolutely stationary. In the warm stillness of a summer evening, for example, or the deep silence of a frozen winter's night, who could possibly believe that the Earth is moving? If the Earth rotates, argued early thinkers such as Hipparchus and Ptolemy, why doesn't its motion produce a spectacular breeze through your hair? In fact, the Earth's atmosphere colludes with gravity in promoting the illusion of a static world. It, too, is bound to the planet, and rotates faithfully with it. The wind in your hair is usually the result of atmospheric weather conditions or, occasionally, from driving in fast cars with the top down. Hipparchus and Ptolemy didn't know that, however.

The Earth's rotation is revealed to us primarily by the apparent movement of the heavens. In reality, this leisurely turning of the sky means that a person standing at the equator has an eastward speed of nearly 1700 km/h – a fair old lick. But there remains some doubt among intelligent people about exactly how this translates into everyday timekeeping – as revealed by the listeners' questions below.

It took until the time of Nicholas Copernicus (1473–1543) for the idea that the Earth moves through space to become widely accepted. Today we know with absolute certainty that the Earth is bounding along in its orbit at some 100,000 km/h, or about 30 km/second. That is a big step up from the rotation speed, as of course it must be, for the Earth has to cover just under a billion kilometres in its annual tour around the Sun.

The planet's speed in its orbit is truly spectacular, but it would be wrong to imagine it streaking through space like a jet aircraft. Whereas a cruising Boeing 747, for example, covers almost four times its own length every second, in relation to its size the Earth travels along in its orbit much more sedately. It takes a full seven minutes for the planet to move through a distance equal to its diameter.

The Earth's orbital mechanisms are complemented by its other interactions with celestial objects. Tides and geomagnetic storms are just two of the common phenomena that have an extraterrestrial origin. While their workings are scientifically well understood, their effects on everyday life often leave unanswered questions in the minds of non-specialists. Fortunately for us all, a few people interested in probing beyond the commonplace have taken the trouble to raise their questions on air. The answers are presented below.

Apart from a handful of astronauts, none of us have ever experienced any environment other than the Earth's biosphere. It is hard to overstate nature's achievement in creating a balance between all the processes at work in that delicate membrane.

Regrettably, even in this enlightened age, it is a balance that we humans seem hell-bent on overturning. Perhaps the illusion of the commonplace on Earth is just too convincing for our own good.

MARKING TIME: EARTH'S ROTATION

How are pictures of circular star trails made?

Images of star trails can be extraordinary, none more so than the well-known photograph of the Anglo-Australian Telescope dome taken by my colleague, David Malin. It shows the imposing observatory building by night against a backdrop of circular trails, which look for all the world like a shooting-gallery target in the sky.

David took this picture by setting up a tripod-mounted camera (the old-fashioned photographic kind, rather than a modern digital), pointing it towards the south and latching open the shutter to make a time-exposure. He set up his equipment early on a moonless winter's night and then went to bed, returning just before dawn some nine-and-a-half hours later to close the shutter and retrieve the camera.

The processed film revealed all the interesting things that occurred while the shutter was open. Most striking, of course, is the dramatic bulls-eye of star trails. The Earth rotated through almost half a turn during the exposure, so the image of each star is smeared out into a near-semicircle. Overlapping images produce the effect of circular trails, and the circles are all centred on the South Pole of the sky – the point about which the whole sky appears to rotate. This is simply the extension of the Earth's axis into space, and it just so happens to be at the same angular height above the horizon as the latitude of the observatory. Rather convenient, really.

If this photo had been taken in the northern hemisphere, it would have shown a bright, almost stationary star image at its centre – the Pole Star, or Polaris. Some of the indigenous peoples of North America call it 'the star that does not walk around', a beautifully evocative name. There is, in fact, a southern pole star, known to astronomers as Sigma Octantis, but it is very faint – almost at the limit of naked-eye visibility – and seems to have had no significance for the Aboriginal peoples of Australia.

Star trail pictures made in colour, like David's famous photograph, prove that the stars are far from uniform in hue. Many are whitish, but there are obvious red and blue trails as well. Astrophysics tells us that the colours indicate the stars' temperatures, from the coolest (reddish-brown) to the hottest (blue). The images also show that the sky itself is faintly luminous, as revealed by dark foreground objects such as trees being silhouetted black against the sky.

Such time-exposures can also show human activity. Bright wavy lines are the flashlights of observers who come out to check the weather during the night. And astronomers' cars occasionally drive by, leaving behind them straight lines from their dimmed parking lights.

In making his star trail images, however, David Malin had to contend with one particularly idiosyncratic human activity. The human in question was a gifted computer programmer by the name of Patrick Wallace, who also had a remarkable talent for practical jokes. If Patrick discovered that one of David's time-exposures was in progress, he would take the dome elevator to the high walkway surrounding the building and, with the biggest, brightest flashlight he could find, trace out in huge letters the rudest word he could think of. And it is quite astonishing how faithfully such optical graffiti are reproduced on photographic film.

The bull's-eye of star trails framing the giant dome of the Anglo-Australian Telescope is dramatic evidence of the Earth's rotation during this all-night photograph. The wavy line near the top of the building is the flashlight of an observer checking the weather from the high walkway. Fortunately, this astronomer resisted the temptation to write rude words with the torchlight. (Anglo-Australian Observatory/David Malin Images.)

What is the precise period of the Earth's rotation in a 'day'?

The short answer to this question is 24 hours, but of course there's a bit more to it than that. If you imagine the Earth rotating on its axis, it's easy to think of a day as the length of time between two successive noons at any given location. (By noon, we mean the time at which the Sun is at its highest in the sky – that is, when it is due South in the northern hemisphere or due North in the southern hemisphere.) We call this period a solar day, and divide it into 24 hours of solar time.

The listener who asked this question had recognised that things aren't quite that straightforward. The complication arises because the Earth isn't just rotating on its axis, but is also moving around the Sun in its orbit. Both these motions take place in the same direction – anticlockwise as seen from above the Earth's North Pole. Thus, during the interval between one noon and the next at any given location (i.e., a solar day), the Earth has to turn through slightly more than a complete rotation to bring the Sun to the same point in the sky, having progressed that bit further along its orbit.

You can define a particular type of day that is exactly one complete rotation of the Earth if you measure time by the stars rather than the Sun. That's because stars are much, much further away, and their positions relative to the Earth change only imperceptibly. So if you define a 'star day' as the time between two successive due-South or due-North passages of the same star – what you might describe as two 'star noons' – then you have the true period of rotation of the Earth. It's called a sidereal day, which is just a fancy way of saying star day. In units of solar time, a sidereal day is 23h 56m 04s long, just a tad shorter than the solar day – exactly as you would expect.

In a direct analogue of solar time, astronomers divide the sidereal day into 24 'sidereal hours' of sidereal time. They use this for convenience when working at their telescopes, as it allows them to keep track of which stars are visible. Of course, unless they're radio astronomers (who don't care), they also have to keep an eye on the solar time, which tells them when the Sun is going to come up.

How do you tell the time with a sundial?

Sundials come in dozens of shapes and sizes, but they all incorporate the principle of telling the time by means of a shadow cast by the Sun. The shadow is formed by a component curiously named a gnomon (pronounced, gnormally, with a silent 'g'). It may be a vertical or inclined spike, a straight edge on a piece of metal, a length of wire, or even a human being – temporarily, of course. Sometimes the 'shadow' is actually a spot of light made by the Sun shining through a hole in something, but the principle remains the same.

The time is read from a numbered dial on which the shadow (or light-spot) falls. Naturally, the dial is calibrated only for the hours of daylight – although in summertime at high latitudes these can extend surprisingly close to midnight. Incidentally, the direction of rotation of the shadow cast by a vertical gnomon in the northern hemisphere is the reason that the hands of a clock go around the way they do. Clockwise.

All that is the easy part. The fiddly bit comes in turning the time shown on the sundial into ordinary clock time – because they aren't the same thing. Depending on the size of the dial, you might be able to read off the time to the nearest minute or two, using the gnomon's shadow like the hour hand of a clock. What you would get is something called 'local apparent solar time'. But if you used this to schedule your appointments, you could be disastrously late – or embarrassingly early.

That is because some subtle corrections have to be applied to local apparent solar time to turn it into ordinary civil time. The first one is the trickiest, and it is necessary because the

Sun runs fast at certain times of the year and slow at others. That sounds bizarre, I know, but it happens for two reasons. First, the Earth's orbit around the Sun is not perfectly circular, so the planet moves a bit more quickly at the beginning and end of the year (when it is closest to the Sun) than it does in June and July. Second, the Earth's axis does not stand exactly upright in its orbit, but is tilted at an angle of 23.5 degrees to the 'vertical'. This tilt gives us the seasons, of course, but it also combines with the Earth's varying speed to affect the rate of local apparent solar time – sundial time.

It was only with the invention of mechanical clocks that people realised how irregular apparent solar time is, and that they needed to adjust it to get standard clock time. This correction is quaintly known as the equation of time, and it varies throughout the year. It is the 'error' of the real Sun compared with a perfectly regular fictitious Sun, and it can be as much as 16 minutes. Only on four days of the year (15 April, 14 June, 2 September and 25 December) does the correction fall to zero.

To get from sundial time to civil time, therefore, the first step is to apply the equation of time for the particular date of the year. This will give you the local *mean* solar time. A very few specially designed sundials make this correction automatically, but for most garden sundials you have to do it yourself. In this day and age, of course, the equation of time is most easily found from the World Wide Web (for example, at www.sundials.co.uk/equation.htm).

The second correction is more straightforward, since for any given place it's always the same. You will have to make it unless you are particularly lucky and your exact location coincides with the defining longitude for your time zone. For example, Eastern Australian Time is defined by the longitude 150 degrees east of Greenwich. Since one hour of time is equivalent to 15 degrees of the Earth's rotation, this gives a time zone ten hours ahead of Greenwich Mean Time. (Greenwich Mean Time, by the way, is just mean solar time at the longitude of Greenwich. It is also called Universal Time.)

If you are reading the time from a sundial in western Sydney, which is at a longitude of 151 degrees east, you are four minutes (i.e., one degree of the Earth's rotation) ahead of the standard time in your time zone. You need to subtract that amount from the local mean solar time you've already established from your sundial – whereupon you'll have reached your goal of standard clock time. *Voilà!* That's it. Well, unless daylight saving is in force – in which case you'll need to add another hour.

Why do we need leap seconds?

The most accurate clocks available today rely not on the swing of a pendulum, nor the oscillation of a balance wheel – nor even the vibrations of a quartz crystal – but on the subtle behaviour of atoms. The best of them are accurate to within one second in a few million years, and define the internationally adopted standard timescale of TAI (Temps atomique international – International Atomic Time in French, in case you're wondering).

Long before such astonishing accuracies were achieved, however, it was recognised that the Earth is not a particularly good timekeeper. Not only does its orbital motion make sundial time seem to go alternately fast and slow (see the previous answer), but the Earth's rotation on its axis is slightly irregular – and is also very gradually slowing down. The main reason for this deceleration is the transfer of rotational energy from the Earth to the Moon, which responds by furtively drifting away from us. (We'll look at this in more detail in Chapter 6.)

The slow-down of the Earth's rotation is almost imperceptible. Our days are, indeed, getting longer – but only by about 0.002 seconds per century. However, even that slight change raises a potential problem for we clock-bound humans on Earth. As the planet's rotation slows, the everyday events that define the passage of time – sunrise, noon, sunset, and so on – begin to get slightly out of step with the relentless ticking of International Atomic Time.

If left unchecked, the cumulative effect of this error would amount to several seconds within a few decades, and the first people to notice would be astronomers and satellite-tracking station personnel trying to point their telescopes in the right direction. After a few hundred years, though, the error would amount to minutes or hours, and no one would be able to ignore it.

In order to avoid this embarrassing situation, the International Earth Rotation and Reference Systems Service at the Paris Observatory decides twice-yearly whether it will add a second to TAI on 30 June or 31 December. Not surprisingly, given the Earth's overall slow-down, there has never been a subtraction, although that could happen if the need ever arose. But as of 2007, 23 leap seconds have been added to TAI since the first one on 30 June 1972. This doesn't mean that our days are now 23 seconds longer than they were then – merely that the clock has been reset by one second 23 times. So far, then, all is well with our clocks, and we won't find sunrise gently slipping towards three o'clock in the afternoon anytime soon.

There is, however, some controversy attached to the issue. The occasional introduction of a leap second is a pain in the neck for anyone writing computer software that has a critical dependency on time. Moreover, technicians operating time-signal satellites (such as the US Global Positioning System, or GPS) have to make tedious adjustments whenever a leap second is added. This has prompted an American delegation to the International Telecommunications Union to request that the difference between solar time (specifically, Coordinated Universal Time, a derivative of Greenwich Mean Time) and

TAI be allowed to range up to a maximum of an hour. At the time of writing, this issue is still under discussion, but if it were implemented, it would mean a major change in our everyday understanding of time.

Finally, talking about atomic clocks reminds me that many years ago, when I was an astronomer at the Royal Greenwich Observatory, I used to chat with members of the Time Department, who were responsible for producing the iconic 'six pips' hourly time signals of Greenwich Mean Time. One day, they were very excited because they had just installed a new atomic clock of incredible accuracy – one part in 100 million million, or one second in three million years.

'Does that mean, then,' I asked one of them, 'that if I come back in three million years' time, your clock will be no more than a second out?'

'Oh, no,' he said. 'These clocks only last for five years.'

Is it possible for an aircraft to keep up with the dawn?

Indeed it is, if the aircraft is flying at a high enough latitude either North or South of the equator. The best way to imagine how this can happen is to remember that every point on the Earth's surface (other than the exact poles) is constantly moving eastwards due to the planet's rotation. Because the planet is a solid body, the actual speed at which you're moving depends on your latitude. In London or Berlin you are doing around 1,000 km/h, in New York or Beijing almost 1,300 km/h, while closer to the equator – in Sydney or Cape Town – you are cruising along at about 1,400 km/h. At the equator itself, your speed is greatest – a cool 1,675 km/h.

Now let's assume for the sake of argument that a commercial jet has an average speed over the ground of 1,000 km/h (which is not too far from the truth). Then, if it flies westwards at a latitude where the Earth's rotational speed is also 1,000 km/h, it will effectively 'stop' the Earth's rotation. If a flight from, say, Scotland to Canada takes off at dawn, it will still be dawn when it lands five or six hours later. Of course, this is an oversimplification, since airspeeds are lower during climb and descent, but the basic principle holds true.

Most such flights would take a route passing closer to the North Pole than their start and finish latitudes (a so-called 'great circle route', which is the shortest distance between any two points on the globe). They could thus overtake the dawn, and that is why it is sometimes possible to land at an earlier local time than you took off. This was especially true in the era of transatlantic Concorde flights. That stunningly beautiful but dreadfully polluting aircraft cruised at well over 2,000 km/h, making this kind of 'time travel' a routine business.

DOING THE ROUNDS: EARTH'S ORBIT AND THE FOUR SEASONS

Why do the dates of the equinoxes and solstices vary from one year to the next?

By definition, the astronomical seasons begin on the dates of the equinoxes and solstices, which are on or about 21 March (vernal equinox), 21 June (summer solstice), 23 September (autumnal equinox) and 22 December (winter solstice). (These dates apply to the northern hemisphere – they are shifted by six months in the south.)

The exact instants of these phenomena are defined by astronomical events. At the equinoxes, the Sun is crossing the celestial equator (which is just an imaginary extension of the Earth's equator into the sky) as it moves northwards in spring or southwards in autumn. (Again, these directions are reversed in the southern hemisphere.) The solstices are at the limits of the Sun's motion, where it seems to stand still at its highest or lowest points in the sky before heading back towards the equator again. Of course, the Sun's annual North–South journey through the sky is simply due to the 23.5-degree tilt of the Earth's axis to the 'vertical' (the direction perpendicular to its orbit). Thus, at the solstices, the Sun is 23.5 degrees North or South of the celestial equator.

The times of the equinoxes and solstices vary from one year to the next principally because a solar year does not contain an exact number of days. In fact, there are 365.242 days in a solar year, and it is the addition of a day in leap years that prevents the calendar getting too far out of step with the seasons. The resulting four-year cycle is what causes the slight variation in the dates of these phenomena from one year to the next.

Incidentally, a common misconception about the seasons is that they are caused by changes in the Earth's distance from the Sun, rather than the tilt of the Earth's axis. The non-circularity of the Earth's orbit amounts to only 3 per cent when you compare its greatest distance from the Sun to its shortest, so it has very little effect on the seasons. Even so, few Australians would be surprised to hear that it is during the first week in January that the Earth passes its closest point to the Sun. Lounging on a beach or trying to keep cool indoors, it is easy to imagine our star's real distance of 146 million kilometres shrinking to a few sweltering miles. It's less easy to come to terms with the same cosmic milestone in Europe or North America, though, as you scrape the frost from your car's windscreen on icy January mornings.

Why does the direction of sunrise or sunset change throughout the year?

The gradual progression of sunrise and sunset along the horizon is very noticeable, and for many people provides visible evidence of the progress of the seasons. It is caused by the Sun's changing daily path through the sky as a result of the 23.5-degree tilt of the Earth's axis to the 'vertical'.

At the equinoxes (around 21 March and 23 September), when day and night are both approximately 12 hours, the Sun rises due East and sets due West. This is true everywhere on Earth, but the effect at other times of the year depends on the latitude of the observer. In Sydney, for example, as the Sun's path gets higher in the sky during springtime, it rises progressively south of east and sets south of west. At the longest day (the summer solstice around 22 December) it is rising ESE and setting WSW. At midwinter, the converse is true – it rises ENE and sets WNW. Throughout the year, the Sun's rising or setting point changes by almost 55 degrees of bearing along the horizon.

At higher southerly or northerly latitudes, the change in bearing is even more pronounced. At the latitude of Edinburgh, the difference between midsummer and midwinter sunrise (or sunset) is as much as 90 degrees. Midsummer sunrise on 21 June is in a north-easterly direction, while it sets in the north-west. The Sun's return path below the northerly horizon at this time of the year is short and shallow, producing those brief, twilight nights that so charm visitors to Scotland in June and July. Pity about the rain and the midges, though. At midwinter, sunrise is to the south-east, followed a few hours later by sunset to the south-west. It's a good thing the Festive Season is there to make up for the long, dark nights in between.

See also: Why does the direction of moonrise or moonset change from night to night? (Chapter 6)

Why are there more than 12 hours of daylight on the day of the equinox?

The term 'equinox' implies that day and night are of equal length, but tables of sunrise and sunset times show that the days when there are exactly 12 hours of daylight don't correspond to the dates of the equinoxes. For example, in London in 2009, the interval between sunrise and sunset is 12 hours on 17 March and 25 September, whereas the equinoxes themselves occur on 20 March and 22 September. At the equinoxes, the Sun appears to be above the horizon for about eleven minutes more than the straight 12 hours.

The reasons behind this go to the heart of how sunrise and sunset times are defined. It is the alignment of the Sun's upper edge, not its centre, with the horizon that marks the moment of sunrise or sunset. The refraction, or bending, of sunlight by the Earth's atmosphere must also be taken into account. The result of these two factors is that at the calculated instant of rising or setting, the true centre of the Sun's disc is almost a degree below the horizontal plane, rather than exactly in it. Thus, the 12-hour days occur *before* the vernal equinox (due to the Sun's northward progression at this time) and *after* the autumnal equinox (due to its motion southwards).

If the Sun's rising and setting times were computed for its centre lying exactly in the horizontal plane, and if atmospheric refraction were ignored, then the 12-hour days would indeed coincide with the equinoxes. In practice, of course, sunrise and sunset times often disagree with predicted values in tables because the visible horizon is shaped by features in the landscape rather than a true horizontal plane.

Why doesn't the shortest day coincide with the latest sunrise and earliest sunset?

Of all the little puzzles that surround the changing of the seasons, this one is the trickiest, and it never fails to impress me how many people notice it. It's to do with the solstices, when the Sun 'stands still' in its gradual progress northwards or southwards in the sky. The Sun's angular height above the horizon at noon is then at its minimum (winter solstice) or maximum (summer solstice).

These solstices coincide with the dates of the shortest and longest days but, paradoxically, don't correspond to the dates when the Sun begins to rise earlier and set later in winter – nor when it begins to rise later and set earlier in the summer. For example, in London in 2009, the earliest sunset is on 12 December, the winter solstice is on 21 December and the latest sunrise is on 30 December. Likewise, in summer, the earliest sunrise is on 17 June, the solstice is on 21 June, but the latest sunset isn't until 25 June.

So why don't these two bewildering sets of dates coincide? The reason has to do with the Sun's bizarre behaviour in running fast and slow, as described in the answer on sundial time. It's the good old equation of time again. At certain times of the year the interval between noons on successive days is slightly greater than 24 hours of clock time, and at other times slightly less, the differences cancelling one another out over the whole year.

Near the summer and winter solstices (June and December), the time between successive noons is respectively 13 and 30 seconds less than 24 hours. These differences are greater than those between the sunrise and sunset times on successive days, so they become the dominant effect. This, in turn, causes the staggering of the dates of the earliest or latest sunrise and sunset times, and the solstices.

The effect is more pronounced in December than June because of the larger discrepancy between the solar day and 24 hours of clock time. It is also less pronounced at higher latitudes, due to a greater daily change in sunrise or set times. And how about this – if you were to plot tables of sunrise and sunset times based on *sundial* time rather than clock time, the effect would disappear altogether.

EARTHLY ATTRACTIONS: MATTERS OF GRAVITY AND MAGNETISM

What would happen if you could drop a stone down a hole through the Earth?

Right. So you get your spade, and start digging. As the hole gets deeper, you line it with super-strength material to stop the walls caving in under the immense pressure exerted by the Earth's soft rocky mantle. As well as being super-strong, this stuff must also be extremely heat-resistant, so that as you go deeper still, the molten iron sloshing around in the Earth's outer core won't melt it. The central inner core is solid enough, but because it's made of iron and nickel your spade will need an industrial-strength blade. And it's getting hotter still – perhaps as much as 7,500 °C. You'll also need radiation protection to guard against the decay products of uranium and other radioisotopes.

Eventually, you'll pass the centre and come up through the corresponding layers on the other side. And when you eventually emerge at the surface, you'll have the satisfaction of knowing that you've dug a hole right through the Earth's diameter. Well done. Now you can perform the experiment this listener had in mind when asking the question. Oh, did I mention that you also have to be a superhero...?

Enough of this rubbish – even if it does give some idea of our planet's complex inner structure, which we shall need later when we discuss its magnetic field. The experiment of dropping a stone down a hole that goes all the way through the Earth can never be carried out, but thanks to the wonders of physics, we know exactly what would happen if it could. And the answer has an appealing simplicity – at least, in the imaginary situation where the Earth has no atmosphere and is perfectly homogeneous inside.

The stone would begin by accelerating at exactly the same rate as if it had been dropped from a high building, gaining 9.8

metres/second (or 35 km/h) of speed for every second of time that passes. It turns out that the gravitational force drawing it to the Earth's centre is proportional to its distance from the centre, which means it will continue to accelerate all the way down. As the stone approaches the centre, the acceleration decreases until, when it gets there, it is briefly zero. By now, however, the stone is cruising at nearly 8 km/second, so it just keeps on going as the Earth's gravity begins to slow it down during the climb back to the surface. Eventually, it gets to the other side, coming to rest at the same moment as it reaches ground level.

Unless somebody grabs it, the stone will then begin to fall again, and will eventually reappear back where it started. Without any air resistance to slow its motion, the stone will keep oscillating like this *ad infinitum*. Each round trip will take 84 minutes 29 seconds, and it turns out that this is exactly the same length of time that a satellite would take to complete an orbit if it was moving just above the Earth's surface. Such a spacecraft – or Earthcraft – would have a constant speed equal to the falling stone's maximum speed, so there is clearly an intimate connection between these two hypothetical modes of travel. The link is the combination of the Earth's gravity and diameter which, of course, are the same in both cases.

Does the Earth wobble on its axis?

The listener who asked this question was thinking of the possible consequences of a catastrophic geophysical event such as the 2004 Boxing Day earthquake off Sumatra, with its resultant tsunami. There are, however, several types of 'wobble' that happen all the time.

The first and most important of these is a regular and predictable phenomenon known as 'precession', which is exactly analogous to the slow rotation of the axis of a spinning top that has not been set up quite vertically when spun. In the Earth's case, the tilt of the axis is 23.5 degrees (sounds familiar?) and its rotation takes some 25,800 years to complete. One of the many consequences of this is that with the gradual passage of time, different stars become the Pole Star (see the answer on star trails in this chapter). It was Hipparchus who discovered this effect, in about 150 BC, and it is due to the combined gravitational pull of the Moon and Sun on the Earth's equatorial bulge – its slightly oversized waist.

Superimposed on the precession of the Earth's axis is another, much smaller effect that was not discovered until the 1740s. This is a slight back-and-forth nodding of the axis, again due to gravitational interactions with the Sun and Moon, known as 'nutation'. There you go – it's so little known that my spell-checker hasn't heard of it. Yet.

Both precession and nutation are movements of the Earth's axis in space, but the axis also moves very slightly in relation to the planet's crust. This so-called polar motion is irregular, and is due to the fact that the Earth's axis does not pass exactly through the planet's centre of mass. That isn't surprising, given that a fair proportion of the planet's interior is in a molten state. The net result is that the respective poles wander around randomly within an area of the Arctic and Antarctic ice some 20 metres across.

Finally, yes, natural disasters such as the Boxing Day earthquake do cause a slight motion, or wobble of the poles relative to the crust. But they are below the level at which measurement is possible. The dreadful event of 2004 had a magnitude of 9.0 on the Richter Scale, enough to wreak devastation on a huge scale. Yet its calculated effect on the position of the pole was only some 27 millimetres, or a little over an inch. And its effect on the length of the day was minuscule – less than 3 microseconds. All of which simply highlights just how fragile and tiny the human environment is compared with the planet itself.

See also: What would be the effect on Earth of the Moon not being there? (Chapter 6)

Why are there two high tides per day?

Most people know that ocean tides are a consequence of the gravitational attraction of the Moon on the Earth. Some people also know that the Sun, too, plays a role in this – in fact, its effect is about one-third that of the Moon. A handful of people realise that tides also occur in the solid crust of the Earth, which can move up and down by as much as 25 centimetres. But hardly anyone seems to understand why there are two high tides each day. Hence this hoary old question.

Tides are caused by the gravitational pull of one celestial body on another, and happen because the force exerted by the attracting body (say the Moon) on the 'attractee' (the Earth) is different depending on where you are standing. Specifically, the force is greatest on the Moonward side of Earth, and least on the far side of the Earth from the Moon. The centre of the Earth feels an intermediate force. Thus, the attracting body (the Moon) is trying to elongate the Earth in its direction, rather than just suck water up towards it. This elongation effectively raises a tidal bulge on both sides of the Earth, which causes two tides per day as the Earth rotates underneath it.

In practice, the timing and height of the tides are very strongly influenced by local topography such as the depth of the sea floor or the presence of narrow straits constricting the flow. Nevertheless, there are always two cycles per day. The highest tides occur when the Sun and Moon are aligned relative to Earth (i.e., near full or new moons). These so-called spring tides also produce the greatest difference between high and low water. Neap tides occur when the Sun and Moon are at right angles (first and last quarter), and are more moderate in their effects. But there are *still* two high tides per day...

Will the Earth's magnetic field reverse?

Even the boffins at the British Geological Survey don't know the answer to this, so you're unlikely to get a categorical one from me. My answer is the same as theirs: a definite 'Maybe'.

A reversal of the magnetic field means that the North and South magnetic poles interchange. You get out of bed one morning, and your compass is suddenly pointing the wrong way. Good thing you didn't need it to find the bus stop. Of course the process is likely to happen much more slowly, perhaps over several thousand years. There may even be dummy reversals, in which the Earth's magnetic field strength weakens over time as if a reversal was approaching, but then regenerates with the original North–South orientation.

What evidence is there that such a reversal does not simply belong in the realm of fiction? The answer lies in the crystalline structure of volcanic rocks, particularly those lying deep under the ocean floor. These rocks faithfully preserve the direction of the Earth's magnetic field, and core samples obtained by deep-ocean drilling reveal the changes that have occurred in the various geological layers. In the recent past – over the last ten million years or so – there have been four or five reversals every million years, although there is no regularity in the spacing of these events. Earlier geological eras seem to show fewer reversals.

The challenge for geophysicists is to understand why magnetic field reversals occur. The magnetic field itself originates in the liquid outer core of the Earth, which lies some 2,900 kilometres below the surface. The deepest 1,200 kilometres or so of the core is solid, as we discovered a few answers ago, when we were digging a hole through it. As the core's major constituent is iron, the whole thing acts like a gigantic dynamo, driven by the rotation of the planet and convection currents in the liquid core. Mathematical models of the dynamo's evolution over time suggest that the central solid core might play an important role in magnetic field reversals. The details are, however, very hazy and certainly don't allow us to predict when a reversal might occur.

There is one hint that perhaps the Earth's magnetic field is, indeed, in the early stages of reversing. This is that for the last two millennia its strength has been gradually declining. While it has been measured continuously only over the past 170 years or so, such unlikely sources as the magnetism preserved in Roman clay pots allow us to make earlier estimates. Some scientists have suggested that in as little as 1,500 years' time, the field strength will be zero. So will a reversal follow that? Maybe.

How does the Sun affect communications and power transmissions on Earth?

The Sun occasionally plays havoc with terrestrial communications and power distribution via space weather (a handy euphemism if ever there was one). The term refers to the turbulence of the Earth's interplanetary environment, as it is bombarded with subatomic particles and magnetic fields emanating from the Sun. Of course, at school we are all taught that the space between the planets is a vacuum. This is essentially true, since the particle densities are extremely low (to wit, about five million particles per cubic metre in the vicinity of the Earth – roughly ten billion billion times fewer than in the atmosphere at sea level). However, the rapid ejection of material from the Sun can energise this 'vacuum' to a very high degree.

Even in its calmest moments, the Sun emits a stream of particles called the solar wind. Despite its faintly comical name, this gusty breeze amounts to a million tonnes of subatomic material per second, travelling at speeds up to a million km/h. It shapes the magnetosphere – the elongated bubble defining the region of the Earth's magnetic influence in space. Periodically, however, the Sun undergoes so-called coronal mass ejections (CMEs), which propel energetic particles and magnetic fields into space at up to 1,200 km/second and generate powerful shock waves in the solar wind. Solar flares, which are violent explosions in the Sun's atmosphere, have a similar effect. The underlying mechanism behind both CMEs and flares is thought to be the sudden release of magnetic stress in the Sun – a kind of gigantic magnetic 'twang'.

The end product of this monumental dumping of energy into the inner Solar System is a geomagnetic storm, which seriously disturbs Earth's magnetosphere. Aurorae (the northern and southern lights) become visible at lower latitudes than normal, and the intensity of the ground-level magnetic field increases dramatically. Such activity tends to occur near the maximum of the 11-year sunspot cycle, and each event typically lasts for a day or two (see 'What are sunspots, and do they have any influence on Earth's climate?' in Chapter 7).

As the Earth's magnetosphere is being bombarded by far more high-energy particles than usual, geomagnetic storms can damage electronic hardware in orbiting spacecraft, and are also hazardous for human space-travellers (and perhaps even airline passengers flying over the Earth's poles). They can also reduce the lifetimes of satellites in low-Earth orbit by increasing the atmospheric drag they experience.

The power and telecommunications outages, however, are caused by strong magnetic fields sweeping over electricity transmission or telephone lines. Moving a wire through a magnetic field causes an electrical current to flow, and the same applies with a stationary wire and a moving field. Very large stray currents can thus be generated by geomagnetic storms, tripping overload equipment and bringing the distribution system to a halt.

Perhaps the most striking example in recent years was in March 1989, when a large geomagnetic storm caused a blackout in Quebec that affected nine million customers. Today, with detailed monitoring of the Sun and its environment now a matter of routine, we can have advance warning of any solar activity likely to have adverse effects on the weather. The space weather, that is.

AND FINALLY...

Why is the Earth called 'Earth'?

Mainly because it is what people have walked on since time immemorial. Had we evolved as creatures of the deep, we might have called our planet 'Water'. Once ancient peoples recognised that our world is a bounded, spherical object rather than an infinitely large flattish thing, it was natural that the idea of 'the earth under your feet' should be transferred to the planet itself.

Among all the planets, ours must have the least glamorous name. But for those of us who spend our lives contemplating far less appealing corners of the Universe, it evokes warm sentiments of home as we consider just what a special place it is. Even so, there are some who complain that our planet's name is simply another word for dirt. And I'm afraid there's no denying that, for the most part, that is exactly how we treat it.

CHAPTER 4

OUT OF THIN AIR

LIGHT AND THE ATMOSPHERE

Our view of the sky from the Earth's surface is profoundly affected by the presence of the atmosphere – both by day and by night. But that crucial mantle of air is astonishingly thin. At its normal cruising altitude of 10 kilometres or so, a jet aircraft is above 75 per cent of the mass of the atmosphere. Most of our weather is generated in that shallow layer beneath it which meteorologists call the troposphere. Here, the Sun's warmth conspires with moisture in the atmosphere to drive the changing patterns of sunshine, cloud and rain. With the weather come light shows, natural pageants ranging in subtlety from the twinkling of stars, through rainbows and brilliant sunsets to dramatic bolts of lightning. And none of these are rare events. On a planetary scale, for example, the Earth's surface receives around 100 lightning strikes every *second*.

Near the upper limit of the troposphere, where it becomes the stratosphere, 8 to 15 kilometres above the Earth's surface (depending on latitude), the air is much more rarefied, but here, too, many remarkable atmospheric phenomena have their origins. This is a realm of temperatures below –40°C, where microscopic ice crystals are suspended in thin air. The prism-like effect of these crystals on sunlight produces a wide range of colourful halos, all of striking beauty.

Higher still, at about 50 kilometres above the Earth's surface, the stratosphere merges into the mesosphere. The temperature of what little air remains falls gradually the higher you go until, at the mesosphere's upper boundary (around 100 kilometres high), it reaches temperatures as low as –173°C – a mere 100 degrees above absolute zero. It is in this truly rarefied zone at the edge of space where meteors – shooting stars – appear.

The various phenomena of the day- and night-time skies have stimulated a keen interest among radio listeners, particularly those who live in the country or near the coast. Here, sky-lore is often a part of everyday life – 'red sky at night, shepherd's delight', for example. It is gratifying that the rather elegant science behind many of these familiar events has stimulated dozens of people to pick up the phone and ask 'Why is it so?'

AS PLAIN AS DAYLIGHT: THE ILLUMINATION OF THE SKY

Why is the sky blue?

Poor old Leonardo da Vinci. As an artist who dealt in light, colour and movement, he wanted to understand how nature produced all these phenomena. And he had a jolly good shot at it, from analysing the tumbling of water over stones in a brook to reasoning why the faintly illuminated lunar disc can sometimes be seen between the horns of the crescent Moon – 'the old Moon in the arms of the new' (see Chapter 6). Thus Leonardo might be described as the first modern scientist; certainly he was one of the most accomplished figures of the Renaissance.

One thing eluded him, however. Why is the sky blue? He got close to the answer in a treatise he penned in 1509 (which is owned today by Bill and Melinda Gates, of Microsoft fame). Here, Leonardo explains that he thinks the blue of the sky is caused by the mixture of white light – from sunlight illuminating the air – with the blackness of space beyond. A good idea, but wait a minute – shouldn't that produce grey? And a fairly dirty grey, at that?

While he was on the right track, Leonardo lacked the understanding that would give him the essential missing ingredient, a subtle optical phenomenon called scattering. All credit to Leonardo, though, because it took 360-odd years, until 1871, before the noble English scientist Lord Rayleigh got to the bottom of it all. Scattering is something that happens whenever light rays strike microscopic particles – including molecules of air. Thus, the flood of sunlight hitting the atmosphere is simply scattered in all directions, which explains why the sky is luminous in daytime.

The reason it is blue, though, is that Rayleigh's theory predicted that blue light is about 3.2 times more likely to be scattered than red light in any interaction with a particle. And that is exactly what we see. While the light of the sky still contains red light, it is overwhelmed by the blue.

This also explains why the Sun often looks red or orange when it is very low in the sky. The rays of light reaching us from the Sun have to pass through a much greater thickness of atmosphere than when they are simply beaming down from overhead. As the blue light is scattered off in all directions, the thicker atmosphere depletes the blue component of the incoming rays very effectively: what remains is red light and hence the reddish tint of the Sun. If there are particles of dust or smoke in the atmosphere, the blue scattering is enhanced still further.

Can you see stars in daylight?

The daytime sky is very luminous, but at the same time transparent. This means that objects brighter than the sky can be seen through it. These include the Sun and Moon, of course, and a few short-lived phenomena such as daylight fireballs (rare), daylight comets (extremely rare) and daylight supernovae (incredibly rare). More about such heavenly ephemera can be found elsewhere in this book.

The list of objects brighter than the sky also includes the planet Venus which, for a few weeks every year or so, is far enough away from the Sun in the sky – and bright enough – to hang like a brilliant but still elusive pinprick of light in the blue. Jupiter, too, as the largest planet, is occasionally bright enough to be seen. But the stars? Only Sirius, the brightest star, has ever been recorded as being visible in daylight hours, and that was shortly before sunset on a day of unusual atmospheric clarity.

Sadly, the idea of stars being visible with the unaided eye in daytime is a myth. And sinking yourself to the bottom of a well won't help either, since it doesn't change the relative brightness of star and sky. What does change that relationship – and can, indeed, help you to see bright stars during daylight – is a telescope. Because telescopes magnify whatever is seen through them, the background sky is spread over a greater area of your eye, which dilutes its intensity. A star, however, remains a point of light no matter how much you magnify it, and the greater collecting area of the telescope's lens or mirror only serves to brighten the star image further. This makes it relatively straightforward to find some stars in daylight with a telescope – if you know where to look.

One word of warning for anyone trying this is to make sure your telescope isn't pointed anywhere near the Sun. The telescope's increased light collecting area also works for heat, and anyone unfortunate enough to look at the Sun through a telescope will be left with an irreparably damaged eye. The same warning applies to the use of binoculars – the only difference being that you will ruin both eyes instead of just one.

Is the night sky perfectly black?

Once the Sun has set, and the atmosphere is shaded by the Earth itself, the air ceases to be luminous. This doesn't happen instantly, of course; a rich array of twilight phenomena can be displayed in a clear sky once the Sun has sunk beneath the horizon. When the now-invisible Sun has reached an angle of 18 degrees below the horizontal, however, none of its light is reaching the atmosphere, so any remaining illumination must come from other sources.

For most people, the night sky will still be quite bright, since most people live in cities, and cities pollute the night-time environment with upward waste light. But even far from centres of population, where there are no artificial sources of light, there is still a range of natural phenomena that cause the sky to glow. The most obvious sources are the Moon and the brighter planets, but there are rarer visitors such as comets or the Aurorae (the polar lights).

Even when those are all accounted for, the night sky is still luminous. The main sources of light are dust particles between the planets lit by the Sun (the so-called zodiacal light); a faint luminosity of the upper atmosphere caused by atoms relaxing after a day of solar illumination (the airglow); light from the visible stars scattered by the atmosphere; and a diffuse background of stars that are too faint to be seen individually. The sum effect is easily visible to the unaided eye: just hold your hand up in front of your face, and it will be silhouetted black against the sky-glow.

CIRCLES IN THE SKY: ATMOSPHERIC OPTICS

Why can you sometimes see a circular rainbow from an aircraft?

Rainbows are caused by sunlight falling on water droplets, usually in the form of rain. Each droplet bends, or refracts the light, separating it into its spectrum colours. In addition, the sunlight is reflected within each drop, so it undergoes a major change in direction. In fact, the light is bent through an angle of 138 degrees, so it is within only 42 degrees (i.e., 180–138 degrees) of being completely reversed. That is why you need to have your back to the Sun to see a rainbow.

The net result of sunlight falling on a large cloud of water droplets is that the observer sees a brightly illuminated arc with a radius of 42 degrees, as measured from the eye. It is the splitting of the light into its spectrum (an effect called 'dispersion') that gives the rainbow its characteristic splash of colour, with red on the outside of the arc and violet on the inside. A fainter, secondary bow can sometimes be seen outside the main rainbow. This is caused by light taking a different path through each raindrop, being reflected not once, but twice. The radius of the secondary bow is 51 degrees, and red is now on the inside. Got all that?

At the arc's centre is an imaginary point directly opposite the Sun – the so-called antisolar point – which is always lower than the horizon, and it is this antisolar point that allows completely circular rainbows to be seen. Imagine first that you're standing on the ground. The Sun is behind you, and a distant cloud of raindrops is in front of you. If the Sun is more than 42 degrees above the horizontal (almost halfway to overhead), you won't see a rainbow at all – it will lie beneath the horizon. This is why rainbows are commonly seen in the early morning or late afternoon. In fact, as the afternoon Sun sinks, rainbows can sometimes be seen rising in the opposite half of the sky – a rather pretty effect. When the Sun is actually on the horizon, the rainbow is an exact semicircle.

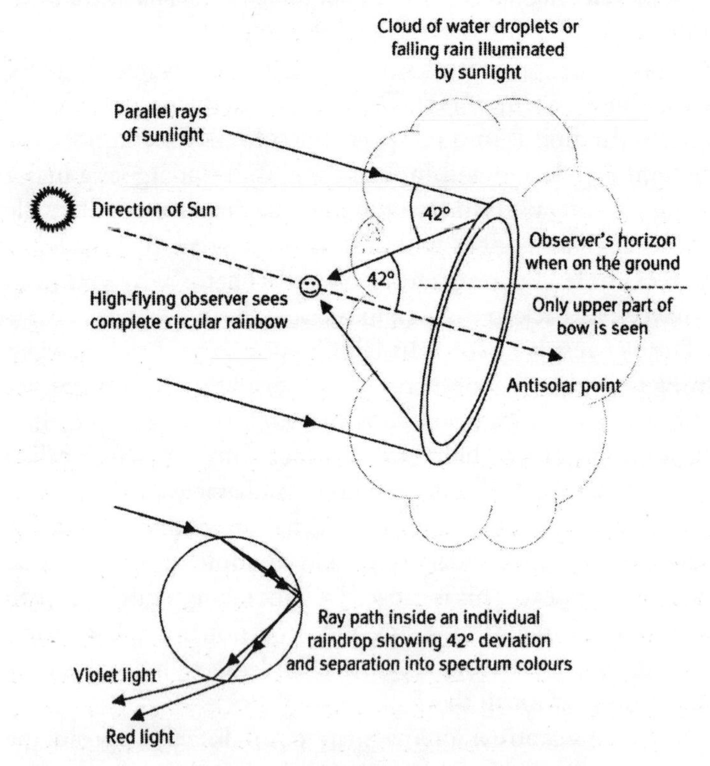

How rainbows are produced. In each individual raindrop, the red rays are bent least, so they form the outer edge of the arc. It all adds up to one of nature's most beautiful spectacles.

Now, as you watch the rainbow, imagine yourself rising wraithlike from the ground. What will happen? Light rays reflected through water droplets below your sightline will start to reach your eyes, so each end of the rainbow will begin to extend. Eventually, enough water droplets will be below your sightline that the two ends will join, and you'll see a complete circle. The effect is, indeed, best seen from an aircraft, but you can simulate it on a sunny day using a garden sprinkler to provide a supply of water droplets between your eye and the ground. It's great fun, but best not done just before you head off in all your finery to that Royal Garden Party.

Can you see rainbows in moonlight?

The full Moon provides about half a million times less light than the Sun, which sounds like a big difference (well, it *is* a big difference), but the human eye has an amazing capacity to adapt to different levels of illumination. Thus, details can be seen easily in a landscape lit only by the full Moon. This sensitivity extends to rainbow phenomena, and when conditions are right (as described in the previous answer), a so-called lunar rainbow can be seen.

It's often said that there isn't enough light in a lunar rainbow for the colours to be visible. The reason given is that the eye receptors triggered by low light levels are different from those used in normal vision and do not provide colour discrimination. Thus, a rainbow is perceived, one that is normal in all respects – except it is white. I know from personal experience that this is not always true, having seen colours in a lunar rainbow on several occasions. But when the phase of the Moon moves away from full, the illumination falls off and the technicolour bow fades to white. Note that this is, indeed, an artefact of human visual perception; if you had sufficiently sensitive equipment and photographed one of these anaemic rainbows, it would look just like a normal rainbow.

Do rainbows form in fog?

Guess what rainbows in fog are called? Yes, fogbows. They are most commonly seen in banks of morning fog when the Sun is still low in the sky. The biggest difference between a fogbow and an ordinary rainbow is that fogbows really *are* white (see the previous answer). They are also slightly broader than rainbows.

The reason for these differences is that the characteristic water droplets in fog are smaller than raindrops. Typically, fog droplets are less than 0.1 millimetres in diameter, and are sometimes very much smaller. It is a well-understood principle in physics that restricting the diameter of any hole through which light is passing spreads out the beam of light. Since each fog droplet acts like a tiny hole, it is the consequent fanning out of the light that causes the greater width of the fogbow, and also makes the various spectrum colours overlap and recombine into white light. Occasionally, the outside edge of the arc is reddish, indicating the presence of droplets at the upper end of the size range.

Fogbows are slightly eerie in appearance, but are well worth looking for on foggy mornings under clear skies, before the Sun burns off the condensation. Go for it this winter.

Why does the Sun or the Moon sometimes have a ring around it?

One of my favourite John Renbourn songs is a traditional number called 'I know my babe'. In it, he tells us that 'She's bound to lerv me some'. ('Lerv'? Oh, well.) But how does he know? 'Why, she throws her arms around me, like a circle round the Sun.' Nice imagery. And, yes, circles around the Sun (and the Moon) are very common.

There are several versions of such circles, all with different characteristics, but the one most often seen is technically known as the '22-degree halo'. What an awful name. But at least it tells you how big it is – 22 degrees in radius, centred exactly on the Sun. Spectrum colours can be seen in the halo (with red on the inner edge), but they are never as vivid as in a rainbow. The halo is normally seen against a background of high cirrus cloud, and is often broken or incomplete. It also forms around the full or near-full Moon. Sometimes the halo can foreshadow rain.

The fact that such halos form in cirrus clouds gives us the best clue to their origin. They are caused not by water droplets, like rainbows, but by ice crystals high in the atmosphere. These crystals are hexagonal in form, rather like a microscopic version of a pencil stub. And the halos they produce are circular for the same reason that rainbows are – light rays passing through many different crystals at the same angle from the Sun trace out a bright circle in the sky. In this case, the crystals have to be randomly orientated to produce the circular symmetry.

As in rainbows, the light's passage through the crystal breaks it up into the colours of the spectrum. This time, however, there is no reflection so the light is only deviated by 22 degrees rather than the 138 degrees of a rainbow.

Another listener asked a related question, which introduces a further aspect of the 22-degree halo: 'I saw a rainbow in the west at 3 p.m., and there was another Sun in it – what was it?' At 3 p.m., the Sun would have been in the western half of the sky, so the 'rainbow' must have been something else (a genuine rainbow would have been towards the east). Most likely, it was part of a 22-degree halo, and the 'second Sun' confirms it. This bright spot was a 'mock Sun' or sundog – a parhelion, to use its rather more elegant technical name.

Parhelia are especially noticeable when the Sun is low in the sky. They look like concentrated, colourful blobs of light in the halo, at the same height above the horizon as the Sun itself. There are often two, one on either side of the real Sun and, like the halo, they are both red on the side nearest the Sun. Sometimes they have horizontal tails stretching away from the Sun. They have a similar origin to the halo itself – ice crystals – except in their case, the crystals are more like tiny hexagonal plates than pencil stubs, and are all floating horizontally like leaves falling from a tree. This uniform orientation concentrates the light in one particular direction and, since the crystals are relatively large, a parhelion is brighter than the halo.

One other striking aspect of parhelia is that as the Sun gets higher in the sky, they appear to detach themselves from the 22-degree halo, moving outwards away from the Sun. They also get fainter. This is due simply to the greater angle at which light is passing through the plate-like crystals.

Finally, and most importantly, who's John Renbourn? He's a legendary British folk-blues guitarist, and a founder member of 'Pentangle'. John's still going strong – check him out on the web.

I saw an upside-down rainbow above the Sun – what was it?

This question threw me completely, but the listener who asked it was a delightful lady who later sent in a photo, confirming the eventual diagnosis of the phenomenon. You would be forgiven for assuming that someone talking about a 'rainbow' *above* the Sun (rather than on the opposite side of the sky) would most likely have seen an isolated part of a 22-degree halo, as explained in the previous answer. But you wouldn't describe that as 'upside-down'. Moreover, the listener was quite clear that the outside edge of the arc was red, and it was violet on the inner edge. In the 22-degree halo, the colours are the other way round.

Stumped, I went back to my books, but quickly found exactly what she had described – a rare halo effect called the circumzenithal arc. The zenith is the point directly above your head, and 'circum' means to go around. This particular arc is never more than a third of a circle in length, always centred on the zenith and sits directly above the Sun. It is at its brightest and most colourful when the Sun is about 22 degrees above the horizon. That figure betrays the origin of the arc, which is again due to light passing through ice crystals.

The photo the listener sent in was a beautiful shot, and looked exactly like the picture in the textbook. She had taken it from suburban Sydney, which just goes to show that you don't have to go to exotic icy locations to see unusual halo phenomena. I've been looking out for one of these arcs ever since – but without success.

TWINKLE, TWINKLE... SCINTILLATING ATMOSPHERICS

Why do stars twinkle?

Once again, scientists have a fancy name for a common phenomenon that everyone is familiar with. It's 'scintillation' and, in this context, it refers to the effect that our continuously moving atmosphere has on light.

If the atmosphere was at exactly the same temperature throughout, the stars wouldn't twinkle and the night sky would lose much of its charm (although, as we'll see, astronomers would be a lot happier). In reality, however, the atmosphere is made up of countless bubbles of warmer and cooler air, jostling each other as they are carried along by the wind. This turbulent motion takes place at different speeds at different heights above the ground, and is often very rapid in the high-altitude jet-stream.

The temperature of air governs its density and, in turn, dictates the extent to which it will bend light, so these moving atmospheric cells act like invisible lenses. They are very weak lenses, because air's refractive (light-bending) power is feeble, but they are strong enough to cause the twinkling of stars – and to compromise observations made with giant telescopes.

Twinkling happens because the moving bubbles of air alternately focus and defocus the starlight arriving at our eyes. The effect is that the star gets rapidly brighter and dimmer – that is, it twinkles, or scintillates. Looking through a telescope, you can detect this focusing and defocusing as the star image seems to contract and expand. The image is also in frenzied random motion, like a moth fluttering around a lamp, as its light rays are bent this way and that in the moving air.

For 300 years, astronomers have referred to the degree of turbulence in the atmosphere as 'seeing'. When the seeing is good, the star images in their telescopes are stable points of light. When it's not, they become inflated trembling blobs, and the exquisite detail that the telescope is capable of revealing is lost altogether in the movement of the air.

Generally, as you look higher in the sky the seeing gets better and the twinkling seen with the unaided eye is reduced, because the path of starlight through the atmosphere is shortest for stars that are overhead. For the same reason, high-altitude sites such as mountain tops offer astronomers better image quality than sea-level locations – there is less turbulent air above them. The seeing is also strongly dependent on local weather conditions. It is often depressingly poor after the passage of a cold front, for example. But the twinkling on such occasions is, well... charming.

I saw a star flashing red, green and blue – what was it?

As we saw in the previous answer, stars low on the horizon are seen through a much greater thickness of air than those overhead. Moreover, the starlight travels through the different layers of the atmosphere very obliquely, which adds a further effect to the normal twinkling or scintillation caused by turbulence. This is dispersion – the splitting of starlight into spectrum colours – and it happens because the steeply angled passage of the light through the air mimics the effect of a glass prism. We have already met dispersion in drops of water and crystals of ice, so it should come as no surprise that it happens in air too. However, air is much less capable of separating out the rainbow colours than water or ice.

If a bright star low in the sky is seen through perfectly steady air, and observed with a high-powered telescope, its light is seen to be drawn out into a short vertical spectrum, with blue and green at the top and red at the bottom. Now imagine the same star viewed through turbulent air, as described above. The moving bubbles of air will bend the coloured rays this way and that, and the unaided eye will see flashes of red, green and blue light. This psychedelic scintillation is most often seen with brilliant white stars; for example Rigel, Vega and, especially, Sirius, the brightest star in the sky. Once again, it is more noticeable from locations near sea-level than at high altitude.

Is it true that planets don't twinkle?

To the unaided eye, planets are indistinguishable from stars. There are five that are visible without a telescope: Mercury, Venus, Mars, Jupiter and Saturn. The rest are too faint, although Uranus sometimes just reaches the limit of naked-

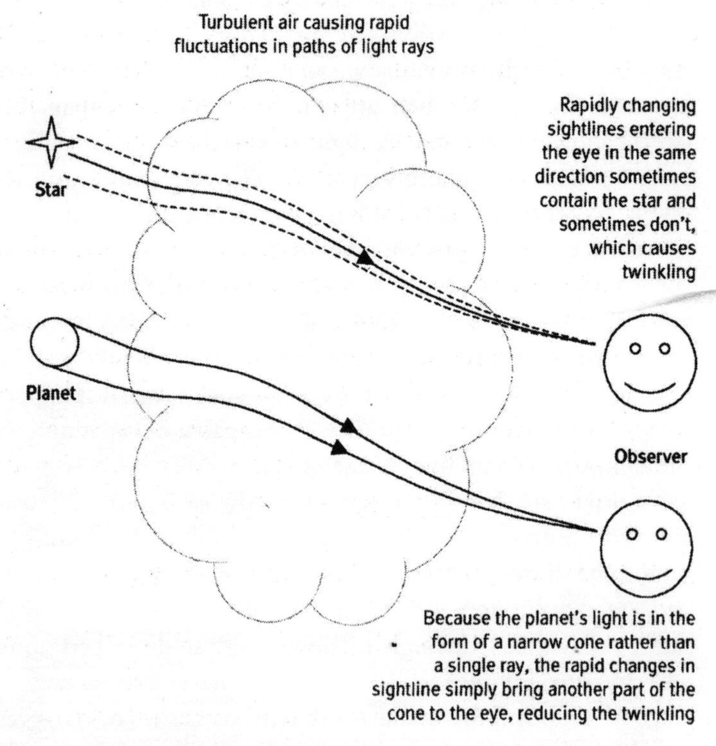

Why planets twinkle less than stars. It's because they are tiny discs rather than single points of light. The effects are very much exaggerated here.

eye visibility. Before the invention of the telescope, the only way people could recognise planets was by their motion among the background stars.

There is a subtle difference between planets and stars that relates to twinkling, however. It is that stars are so far away that, unlike planets, they are effectively single points of light. Planets have measurable diameters, and while in each case it is too small for us to make them out as discs with the unaided eye, it does affect the way they twinkle.

It's easy to imagine a single ray of light from a star being bent by the pockets of air in the atmosphere, causing the starlight to flicker in intensity – or twinkle – when seen with the unaided eye. But when a planet is observed, the eye (albeit imperceptibly) sees a narrow cone of rays, and the bending of these rays just brings another bit of the cone to the eye. The twinkling, therefore, is considerably reduced.

The effect is strongest with the largest planet, Jupiter, whose angular diameter is never less than 30 seconds of arc (a second of arc being 1/3,600 of a degree), and can exceed 50 seconds. Venus can appear even bigger. These planets tend not to twinkle. Mercury and Mars can be much smaller, however, and frequently twinkle away cheerfully when low in the sky – albeit to a lesser extent than any star would. Thus, the rule that planets don't twinkle is not a reliable one.

SPACE INVADERS: SHOOTING STARS AND METEORITES

What are shooting stars (or falling stars)?

Spectacular though they often are, shooting stars are at the smaller end of the cosmic size range. Well, to be honest, they're at the minuscule end. They are particles of dust or tiny stones seldom bigger than an orange pip that hit the upper reaches of the Earth's atmosphere at high speed. The heat generated vaporises them instantly, they shine brilliantly for a few tenths of a second, then they are gone.

These objects have their own (faintly comical) technical vocabulary. While they're floating in space, minding their own business, they are known as 'meteoroids'. When they hit the atmosphere and burn up, they become 'meteors'. And the bigger ones that make it to the ground are 'meteorites'. All very neat and tidy.

Since most people see meteors only occasionally, it often comes as a surprise to learn that up to a billion of them hit the Earth's atmosphere every day, representing a daily influx of around 50 tonnes of meteoroidal material. That gives some impression of the dusty environment in which the Earth orbits the Sun. Depending on where they originate, meteoroids can hit the Earth's upper atmosphere at speeds of anywhere between 11 and 72 km/second, so it's easy to imagine the temperature soaring instantaneously to several thousand degrees as they shoot into the rarefied air. The meteors we see are usually about 95 kilometres above the surface.

The meteorites that make it to the ground, and are subsequently recovered, are valuable samples of extraterrestrial material and much can be learned from their analysis. There are several different categories of meteorite, but about 5 per cent of them contain large amounts of iron. The remainder are stony meteorites.

Some years ago, the United Kingdom Schmidt Telescope at Coonabarabran in north-western New South Wales took a photograph of the sky that included a large galaxy – an aggregation of hundreds of billions of stars deep in space. But the image was ruined (or perhaps immortalised) by a bright meteor that had streaked through the telescope's field of view during the hour-long exposure. It seemed to pierce the heart of the galaxy, highlighting the fact that our view of the sky ranges from the smallest and nearest cosmic objects to the largest and most distant – sometimes simultaneously.

What's a meteor shower or storm?

Meteors come in two basic varieties. Sporadic and shower. You probably knew that already. Sporadic meteors are just particles of dust from the Earth's environment, swept up by the planet as it orbits the Sun. The dust is the last vestige of a disc of material that went into building the planets shortly after the Sun's formation some 4.56 billion years ago.

The best time to look for sporadic meteors is after midnight. During the second half of the night, you are on the side of the Earth facing forwards – that is, into the direction of orbital motion. The meteors then strike the atmosphere rather like raindrops hitting the windscreen of a moving car. Like raindrops, too, they can come from pretty well any direction in the sky.

Shower meteors have a different origin, and behave rather differently. They are associated with comets – those dusty snowballs that orbit the Sun in very elongated paths, often throwing off bright tails of dust and gas as they pass close to our star. Cometary orbits are littered with dusty debris left behind as the comets' icy mantles gradually evaporate. During the Earth's annual tour around the Sun, the planet passes through many of these orbits. Each time it encounters one, the comet's dusty refuse enters the atmosphere from a particular direction, as determined by the Earth's motion relative to that of the particles. That direction remains the same for each shower.

An observer on the ground will see many meteors, all seeming to come from a particular point in the sky. Rather evocatively, this point is called the radiant, and is exactly analogous to the distant convergent point of a set of railway tracks. The tracks are parallel in reality, but perspective makes them look as though they join in the distance.

The various meteor showers the Earth encounters annually are named for the direction from which they appear to come, specifically the constellation that contains the radiant. Don't laugh, but the suffix '-ids' is always added to the constellation name to signify the meteor shower. Thus the Geminids appear to come from Gemini, the Orionids appear to come from Orion, and the Leonids are from Russia (or, more accurately, from Leo).

Some of the more impressive meteor showers, such as the Geminids in early December, will produce around 100 meteors per hour (1–2 per minute). They may flash across any part of the sky, but they all appear to come from the one radiant – a very noticeable aspect. (Detailed information about meteor showers can be found from websites such as that of the American Meteor Society.)

Sometimes, the Earth will pass through a particularly dense part of a comet's dust trail, and much greater numbers of meteors are seen. For example, every 33 years the Earth passes through the orbit of Comet Tempel-Tuttle when the comet is close to the Sun. For a few years either side of that event, enhanced activity may be seen in the November Leonids, which originate from Tempel-Tuttle. This activity can reach truly spectacular levels, such as in 1966, when meteors were seen at the rate of 140,000 per hour (40 per second). Such incredible – but completely harmless – displays are called meteor storms.

What's the difference between an asteroid and a meteorite?

Technically, the question should be, what's the difference between an asteroid and a meteor*oid*, since a meteor*ite* has landed already. That's today's contribution from pedant's corner. However, the question is well put, because things that are large enough to penetrate to the ground (i.e., meteorites) are not tiny particles of dust but larger rocks, most of which have originated in asteroids. A very small proportion of them were ejected from the Moon and the planet Mars as a result of much larger asteroids impacting on their surfaces long ago. (About 30 meteorites are known from each – see Chapter 7.)

The bottom line, though, in this question is that the distinction is a matter of size, and the boundary is rather blurred. For example, a space rock with a diameter of 10 metres could be either a large meteoroid or a very small asteroid. Any object bigger than this usually would be classified as an asteroid.

What's a fireball, and can you see one during the day?

A fireball is simply a bright meteor, usually taken to be one that exceeds the brightness of Venus. About 0.1 per cent of meteors, a relatively large fraction, fit into this category. Fireballs occur when the incoming object is rather bigger than the usual grain of dust or orange seed, perhaps getting on to be the size of a golf ball. Frequently, they are bright enough to illuminate the entire nocturnal landscape and, yes, occasionally they can be seen in daylight. People who have seen fireballs often comment on their striking green colour, a result of oxygen atoms in the upper atmosphere being whipped into a frenzy of excitement by the sudden input of energy. More rarely, red or blue fireballs can be seen.

Fireballs sometimes break up or explode during their entry into the atmosphere, and are then known as bolides. If you see one, it is worth listening out for a few minutes after the event, as they are often accompanied by a sonic boom, which takes a while to make its way down from the upper atmosphere. The second of the two events described below was followed five minutes later by the low rumbling of the distant sonic boom.

Since fireballs are relatively common, most astronomers encounter several throughout their working lives. That's certainly true for me, but two in particular stick in my mind. They occurred within a few months of each other in the mid-1990s.

The first happened one night as I was sitting at home in the dark – probably trying to encourage my baby son, Will, to go to sleep. The curtains were open, and cars occasionally drove past, throwing moving rectangles of light across the living-room walls. Then, a moving rectangle of light tracked rather more quickly across the *floor*... It took a minute or two for me to register that this was a very unusual variety of car – and sure enough, the observatory phones ran hot the following day with reports of a bright fireball.

The other occasion was even more memorable. I was lying on the grass in the backyard with my other son, James – then a toddler – waiting for the planet Jupiter to appear in the summer evening twilight. The sky was still bright, the Sun not long having set, and the one who spotted Jupiter first would win the game. Suddenly, we were amazed to see a dazzling fireball shoot from West to East across the sky. 'Wow – did you see that?' I asked. 'Yes,' said James. 'Do it again.'

CHAPTER 5

WOULD-BE SPACEFARERS

HUMANKIND TACKLES THE FINAL FRONTIER

1957. What a year. The first International Geophysical Year, coinciding with the maximum of the Sun's 11-year activity cycle. The first full year of *New Scientist* magazine – still going strong today. The first broadcast of Sir Patrick Moore's world-famous astronomy TV programme, 'The Sky at Night', also still going strong. And, in the *Eagle* comic, that true-blue space hero, Colonel Dan Dare, was fighting it out with the dreaded Mekon – and outwitting him every time. Science reigned supreme.

In the United States, the Navy was confident that its Project Vanguard would soon result in the first artificial satellite orbiting the Earth. But progress was hampered by rivalries with competing Army and Air Force projects, and several tests failed spectacularly. In the midst of the ensuing bickering, the unthinkable happened. The Space Age took off on the other side of the world – without anyone bothering to invite the United States. It caught a lot of people by surprise.

An 85-kilogram metallic sphere called *Sputnik* ('Fellow Traveller') had been launched into orbit by a Soviet rocket on 4 October, becoming the first human-made object to circle the Earth. It orbited every 96 minutes, coincidentally staying aloft for the same number of days, mocking the western world with its insistent 'beep-beep' signal. Adding insult to injury, on 3 November the USSR followed this feat with the launch of *Sputnik 2*, which carried the first living creature ever to orbit the Earth. It was an unfortunate dog by the name of Laika who took the honours. She survived for a while, but didn't return home. At least, not in the normal way.

The gauntlet was thrown down, and today the Cold War space race that followed is the stuff of legend. Perhaps, as is often said, it spared the world a Hot War of a vastly more serious kind. In any event, the years that followed were bumper ones for science. That rich harvest was heralded by the first successful US satellite, launched, as it turned out, by the Army on 1 February 1958. *Explorer 1* carried Geiger counters that detected radiation belts surrounding the Earth. Even more significant for the future, however, was the end of American inter-service rivalry in space with the formation on 1 October 1958 of the National Aeronautics and Space Administration, more commonly known as NASA. The rest, as they say, is history.

NASA quickly began working on human spaceflight and, following President Kennedy's famous 1961 announcement of the Americans' intention to send astronauts to the Moon, carried out the successful *Mercury* and *Gemini* Earth-orbit programmes. These essential precursors to long-distance human spaceflight paved the way for six *Apollo* lunar landings from 1969 (*Apollo*

11) to 1972 (*Apollo 17*), which effectively brought the space race to a close. As political pressure on the world's scientists eased, the pace slowed, but *Skylab* (the first US space station) was launched in 1973, followed two years later by the joint USA–USSR *Apollo-Soyuz* Test Project. NASA's new Space Transportation System, the Space Shuttle, entered service in 1981, while the Russians launched their long-lived *Mir* space station in 1986, building on their experiences with the earlier *Salyut* project. Today, all these various endeavours have their culmination in the International Space Station, whose mixed fortunes detract little from its enormous value as a genuine multinational venture in space.

These extraordinary achievements have not been without cost, of course, and no one who viewed the television footage will ever forget the loss of the Space Shuttles *Challenger* and *Columbia* in 1986 and 2003. The 14 astronauts who died in those two dreadful accidents bring to a total of 21 the number of men and women who have lost their lives in human spaceflight. Tragic though the losses are, it is a remarkable testament to the care exercised in an inherently dangerous business that they have not been much, much higher.

With the three remaining operational Space Shuttles (*Discovery*, *Atlantis* and *Endeavour*) due to be retired in 2010, NASA has announced its replacement space transportation system. Anticipating a permanent human presence on the Moon by 2024, and crewed missions to Mars further down the track, NASA has unveiled a back-to-the-future design that looks a lot like a scaled-up version of *Apollo*. Known as *Orion*, this project will attract international participation and take humankind on the next leg of its exploration of our interplanetary environment.

A rather different slant on such heroic ventures has recently emerged, and that is the commercial exploitation of human spaceflight. By mid 2007, five fare-paying passengers had had excursions to the International Space Station, each filling an empty seat on a *Soyuz* servicing mission at a cost of some US$20 million apiece. More will follow. But the development by entrepreneur Burt Rutan of space planes capable of suborbital flight has heralded an era in which a space experience – albeit a brief one – might be available to less well-heeled adventurers. Space tourism appears set to rise in popularity over the next decade or so, bringing a completely new dimension to the travel industry. One day, we will have tourist hotels in orbit, providing unparalleled views of our planet with a helping of weightlessness thrown in.

While human space exploration has truly long-term aspirations, by far the greater immediate scientific return has come from projects involving unmanned spacecraft. Since the first data on the Earth's environment were returned by *Explorer 1* in 1958, the unique attributes of an observation platform in orbit have been widely exploited. Information on everything from land-use to climate change, from atmospheric pollution to gravity, has been brought to Earth from space.

And the use of orbiting spacecraft for communications, navigation and weather forecasting really needs no introduction here.

It is in planetary exploration that robotic space missions have truly come into their own. NASA planetary probes like the *Pioneer* and *Mariner* series have been complemented by targeted missions such as *Viking* (Mars orbiter-landers, 1975–76), *Voyager* (outer Solar System, launched 1977) and *Galileo* (Jupiter and its moons, 1989–2002). The Soviet *Venera* series (1961–83) remain the only spacecraft to have landed on the inhospitable planet Venus.

At the time of writing (2007), Mars is a focus of activity, with three operational orbiting spacecraft (NASA's *Mars Odyssey* and *Mars Reconnaissance Orbiter*, and the European Space Agency's (ESA) *Mars Express*) and two incredibly successful

NASA rovers on the surface, *Spirit* and *Opportunity*. Saturn, too, has yielded many of its secrets – particularly on its ring system and satellites – to *Cassini*, a joint venture of NASA, ESA and the Italian Space Agency. Other probes are winging their way across the Solar System, most notably, perhaps, the *New Horizons* spacecraft, which will rendezvous with the dwarf planet Pluto in 2015.

Finally, the ability to place large telescopes above the Earth's turbulent atmosphere gives astronomers a unique (though expensive) window on the Universe. Most of us have seen dazzling images from the 2.4-metre diameter Hubble Space Telescope at some time. That amazing machine has been at work since 1990, and its successor – the James Webb Space Telescope – will go into operation after the Hubble project comes to an end in 2010. Hubble sees the Universe in visible light, but other orbiting telescopes look at regions of the spectrum that are absorbed by Earth's atmosphere and hence are undetectable from the ground. Among the more notable are NASA's Chandra and Spitzer Space Observatories, which survey the Universe in X-rays and infrared light respectively. Each of these projects is yielding spectacular results that could not have been obtained by any other means.

It is an enormous tribute to the publicity departments of the world's space agencies that most of the projects mentioned above are household names, and that the average person in the street has at least a passing acquaintance with current space missions. No doubt that is why radio listeners' questions on space tend to be rather generic in nature. Since most people have heard of the Hubble Telescope, for example, few questions are asked about the actual spacecraft. The technology that keeps it in space, however, is not usually covered in the press releases but is still of interest to listeners. So perhaps that's why they turn to me!

STAYING ALOFT: THE MECHANICS OF SPACEFLIGHT

How do satellites stay up in space?

Most of us are fairly law-abiding when it comes to gravity. Satellites, however, appear to ignore the law of gravity with impunity, sailing gaily through space while all logic says they should fall to the ground. The curious thing is that in fact, they *are* falling to the ground – all the time. Yet they never reach it. Why? Satellites have a spectacularly rapid forward speed, and so, by courtesy of the curvature of the Earth, the ground is falling away from them at the same rate as they are falling towards it. Thus, they just keep on going in everlasting circles. A neat trick, first spotted in the 1680s by a chap called Isaac Newton.

The first requirement for a satellite to stay up in space, therefore, is a very high forward speed; in fact some 8 km/second is required for a low-Earth orbit a hundred kilometres or so above the surface. If an object travelling through the Earth's atmosphere maintained such a speed, friction between the object and the air would heat it to disastrous temperatures and it would burn up – exactly as happens to re-entering space junk. That leads to the second requirement for a satellite to stay in space, which is height. As we saw in the last chapter, there is precious little air above 100 kilometres, but variations in atmospheric pressure and other factors, such as solar activity, mean that for an orbit to be stable, it must be at least 160 kilometres high. A satellite in such an orbit will make a complete revolution of the Earth in about 88 minutes.

This combination of requirements – height and forward speed – create the textbook image of a spacecraft launch: a rocket rising vertically at first, but gradually pitching over to assume a more horizontal position as it gains height. Once in orbit the rocket is turned off, and the delicate balance between Earth's gravity and the satellite's forward motion keep it there. In nearby space, a few hundred kilometres above the surface, the pull of gravity is only slightly less than it is on the ground. This comes as a surprise to most people, who are used to TV images of astronauts surrounded by the weightless detritus of their trade – floating pens, cameras, food capsules, sick bags, and so on. But the weightlessness comes not from a lack of gravity, but from the fact that the spacecraft and its contents are in a state of continuous free-fall.

Another surprising facet of orbital space-flight is that the height of the satellite's orbit is determined by its speed. Slower speeds correspond to higher orbits where the pull of gravity is less, but to attain those higher orbits a spacecraft has to be given additional energy. Thus, if you boost the speed of your satellite by turning the rocket on again, you will gain height. Paradoxically, however, although you have experienced a net gain in energy, your average orbital speed will decrease, so the higher the orbit, the lower the average speed.

Likewise, if your satellite slows down (usually because of the slight drag imparted by the last vestiges of Earth's atmosphere) you will lose height – and speed up. It is this drag that causes the orbit of a satellite to 'decay' and, just as with tooth decay, if you don't correct it, you'll lose it. In this situation, the satellite's decreasing energy causes it to fall deeper into the atmosphere; this brakes it further, and eventually it will burn up in the blazing heat of re-entry. (Not too many teeth go that way, thankfully.)

There's one further trick that can be used to assist a satellite gain a high forward speed at lift-off, and that is to claim a small free gift offered by the Earth itself. If a space launch is made in an easterly direction, the ground speed due to the Earth's rotation can be added to the rocket's forward motion as it accelerates to attain orbital velocity. As we saw in Chapter 3, when we were attempting to keep up with the dawn, every point on the Earth's equator is moving eastwards at about 1675 km/h. Thus, if you launch your satellite on or near the equator, you gain a free 0.5 km/second of velocity, which can produce significant savings in fuel at lift-off. And, with the pump-price of rocket fuel being what it is today, any saving is worth having.

See also: How do you navigate in space?

Do satellites change direction?

The listener who asked this question thought they had seen a satellite abruptly change direction in the course of its passage across the sky, and naturally wondered whether a satellite can be steered around like an aircraft. The answer is yes and no – but mostly no.

Imagine the orbiting spacecraft we were discussing in the previous question. If you could turn the rocket exhaust so that it was pointing at right angles to the spacecraft's direction of motion (though still in the horizontal plane) and then turned the rocket on, what would happen? Well, the spacecraft would change direction. The sideways effect of the exhaust would be to slightly shift the imaginary plane in which the orbit is located. But orbits are *very* big things – bigger than the Earth itself – so to make a significant change in the orbit plane requires a huge amount of energy. That is why once a satellite has been launched into a particular orbit, it is very difficult to change anything except its height.

It's much more likely that this listener had seen something else – an aircraft, or even a weather balloon. It's also possible that they had seen two satellites close together in the sky. If the satellites were at different heights, it would be perfectly possible for the lower of the two to 'disappear' as it entered the Earth's shadow, leaving the other one still illuminated by the Sun. This perhaps would give the illusion of a single spacecraft changing direction. While I was explaining this on-air, a surprising call came in from another listener whose pastime was... night kite-flying. This involves kites fitted with LED lamps, and I had never heard of it. So, yes, an illuminated night-flying kite might just be another explanation for the phenomenon. But whatever it was, it was *not* a satellite swerving round a corner.

Without air to push against, how do thrusters change the direction of a spacecraft?

Ah, nostalgia isn't what it used to be... This is one of the hoary old favourites of the early Space Age, but it's no less interesting for that. When a rocket fires, the exhaust emerges from a nozzle at very high speed, due to the pressure of burning gases inside what is usually known as the combustion chamber. If you imagine the chamber as a sphere with a hole for the nozzle, then you can see that the force on the chamber wall is everywhere balanced by an equal force on the opposite side of it – except directly opposite the nozzle. It's that unequal force pushing away from the nozzle that drives a rocket forward, not the exhaust gas pushing against the air.

When a rocket motor is operated in the atmosphere the exhaust gases are slowed by their interaction with the air, so the efficiency of the motor is reduced. It is in the vacuum of space that they are at their most effective. This applies not only to the spacecraft's main engines, used to change orbital height, but also to the small 'vernier' rockets used for fine manoeuvres.

How do you navigate in space?

Space navigation is actually more straightforward than finding your way around on Earth. That's partly because there are fewer things to bump into, but it's also to do with the way gravity acts to form invisible, preordained pathways through space.

It's a matter of routine for the various tracking stations located on the Earth's surface to get a fix on a spacecraft's three-dimensional position by means of radio signals. If you know where your spacecraft is at any given moment, and can also determine its speed and direction of travel, then you have everything you need to work out where it's going.

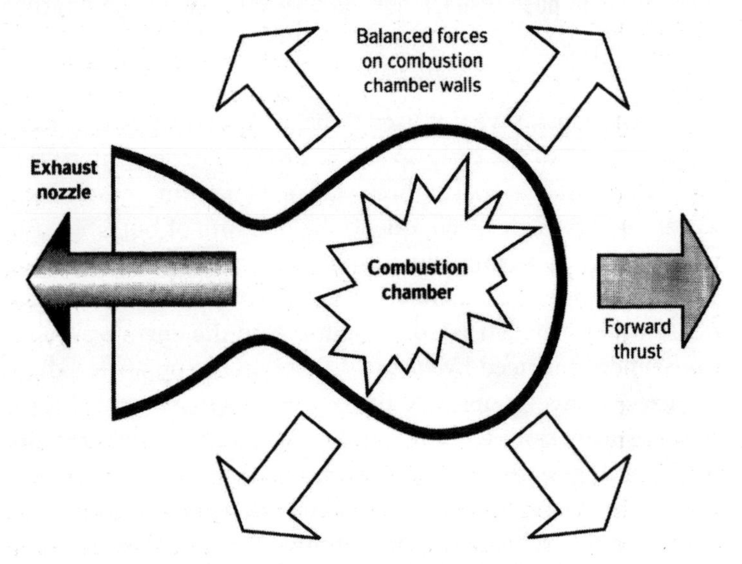

Layout of a rocket engine, showing how the unbalanced force opposite the exhaust nozzle produces a forward thrust. The bell-mouth shape of the nozzle increases the exhaust gas velocity. Just rocket science . . .

That's not because spacecraft move in straight lines – they don't – but because the pull of gravity at *any* point in space is known with great precision. In the simplest case of a satellite orbiting the Earth, the pull is centripetal – that is, directed towards the centre of the Earth. So, if you know the satellite's position and velocity at a particular time, you can then calculate the characteristics of its orbit (for example, whether it's circular or elongated) and hence determine its future position at any point in time.

There are certain effects (usually known by the somewhat perturbing name of 'non-gravitational perturbations') that can mess up this neat and tidy picture. The braking action of the rarefied upper atmosphere is perhaps the most common, but is also well understood and can be allowed for in the calculations. Likewise, the turbulent stream of subatomic particles that we call the solar wind (see Chapter 3) can also be taken into account.

If you want to navigate to another celestial object such as the Moon, the situation becomes a tad more complex but the same basic principles apply. At any point in space, the gravitational pull is a combination of *two* attractive forces, in this case one due to the Earth and the other to the Moon. Of course, the relative intensity of these forces changes as the spacecraft moves along its course, but that intensity is still known with high precision, pre-determining the spacecraft's route. Baby-boomers may recall that the trajectory taken by the *Apollo* astronauts on their way to the Moon and back was actually a figure-of-eight, one of the possible solutions to the complex equations determining these so-called 'three-body orbits' (the third body being the spacecraft).

In such lengthy trajectories, the spacecraft's path is determined by an initial 'orbit injection burn' of the main rocket motor. If necessary, the path can be changed slightly by a mid-course correction, in which the rocket is fired to compensate for any error in the original trajectory.

For travel between the planets, mission planners have to take into account the gravitational attractions of the Earth and the destination planet, together with those of the Sun, the Moon (while the spacecraft is close to the Earth) and, often, the largest planet, Jupiter. In fact, many interplanetary space probes use intermediate planets as staging posts, gaining energy from each encounter in a slingshot fly-by known in the trade as a 'gravity assist'. Sometimes, multiple assists are used, as in the complex seven-year trajectory of NASA's current *Messenger* probe to the planet Mercury (which it is scheduled to reach in 2011).

Modern supercomputers allow us to look at the problem of interplanetary navigation in a rather more refined way. The varying positions of the planets relative to one another, along with the subtle combinations of their gravitational attractions, define an invisible network of constantly shifting pathways through space. These pathways are trajectories along which a spacecraft can move with the minimum expenditure of energy, and the network is sometimes called the 'interplanetary superhighway'.

This futuristic concept is already playing its part in navigating spacecraft to the far-flung reaches of the Solar System.

What's so different about geostationary satellites, and how high do they have to be?

If ever you come across a copy of the February 1945 issue of *Wireless World* – and you never know when you might – have a look at page 58. There, you will find a short letter from a chap called Arthur C. Clarke (later *Sir* Arthur) about the future uses of wartime rockets for space research. In a throwaway paragraph at the end, he suggests that:

> An 'artificial satellite' at the correct distance from the Earth would make one revolution every 24 hours; i.e. it would remain stationary above the same spot... I'm afraid this isn't

> going to be of the slightest use to our post-war planners, but I think it is the *ultimate* solution to the problem [of long-range radio transmissions].

This radical and original suggestion (which Clarke followed up with a brilliantly foresighted article in the October issue) forms the basis of modern mass communications. The only thing he got wrong was the timescale. He expected it to take place 'perhaps half a century ahead', whereas the first geostationary communications satellite (*Early Bird*) was launched just 20 years later, in April 1965. Today, there are literally hundreds of them.

The main point to grasp in understanding geostationary satellites is that normally there is no fixed relationship between an orbiting spacecraft and the Earth's surface. A satellite in a particular orbit zooms endlessly around in that orbit, while the Earth turns ponderously beneath it in an independent fashion. As we have seen, a satellite in low-Earth orbit takes about 90 minutes to circumnavigate the globe. Since the satellite's orbit is usually tilted at an angle to the equator (sometimes quite steeply), successive overhead passages appear to step from East to West across the Earth's surface – as a result of the planet's rotation in the opposite direction. This is how reconnaissance and mapping satellites are able to cover such a large portion of the Earth's surface.

The higher the satellite flies, the longer it takes to complete an orbit, so the successive overhead passages become further apart. Eventually, there will be a height at which the Earth will complete a whole turn during one orbit of the satellite, so each overhead passage occurs at the same location. This is what's called a geosynchronous orbit. Seen from the Earth's surface, the satellite will appear to make a daily excursion up and down in the sky through an angle equal to twice the tilt of its orbit to the equator.

Now, if you launch a geosynchronous satellite into an orbit with *zero* tilt – that is, around the Earth's equator – then even that daily excursion is eliminated, and the satellite will appear to remain permanently above a particular point on the equator. This is a so-called geostationary orbit, and is exactly what Clarke had in mind back in 1945. (In fact, it is sometimes known as the Clarke Orbit.) To be completely accurate, a geostationary orbit is one whose period of revolution matches the rotation period of the Earth as measured by the stars, not by the Sun. (We saw in Chapter 3 that this is called a sidereal day, and is some four minutes shorter than a normal solar day.)

It turns out that a spacecraft is geostationary at a distance of about 42,200 kilometres from the Earth's centre, or 35,800 kilometres above the surface. At that height, its orbital speed is a shade more than 3 km/second. 'Oh, that's easy,' you're probably thinking. 'Much less than the 8 km/second of low-Earth orbit...' But to reach a geostationary orbit, you first have to achieve a low-Earth orbit, then add huge amounts of energy (by firing your rockets) to increase your height until you reach 35,800 kilometres. It took *Early Bird* some 63 hours of repeated rocket burns to attain that first geostationary orbit.

The point of all this, of course, is to allow a communications satellite to 'see' almost a whole hemisphere of the Earth simultaneously, although the equipment carried by these spacecraft usually is targeted to cover a particular footprint on the Earth's surface for broadcasts or communications. Arthur C. Clarke's 1945 vision was amazingly complete, extending not only to the idea of using solar energy to power the satellites, but to details of the antennas that would be needed to make them work properly. Here, too, he was right on the money: 'Small parabolas about a foot in diameter would be used for receiving at the Earth end...' Remarkable.

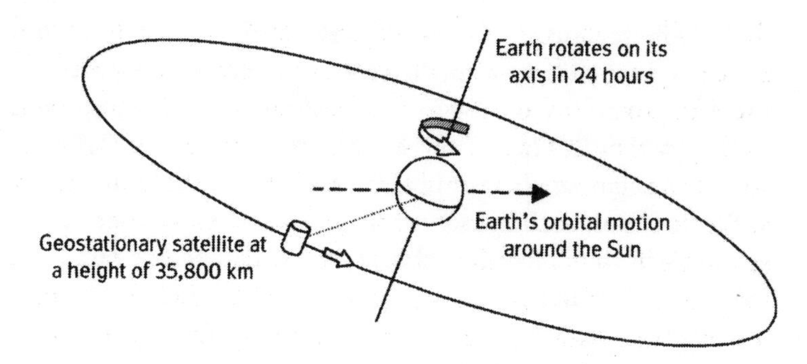

The satellite orbits the planet once in 24 hours, always staying above the same point on the equator

How to make a geostationary satellite stay above the same point on Earth. Its orbit must be in the same plane as the Earth's equator, and its orbital period has to be 24 hours. Satellites in low-Earth orbit (less than 1000 km) revolve around the planet much more briskly.

SEEING THE SIGHTS: VISTAS FROM EARTH AND SPACE

How do we see satellites?

Watching satellites crisscross the night sky for an hour or two after sunset (or before sunrise) is one of the gentler pastimes of the Space Age, and certainly something that was denied to our forebears. The light we see from the satellite is, of course, just reflected sunlight from a Sun still above the satellite's local horizon. The higher the satellite above the Earth, the longer after sunset (or before sunrise) it can be seen – and, incidentally, the longer it takes to cross the sky.

Can you see geostationary satellites with binoculars?

Under normal circumstances, there's not a chance. Because we see them from such a great distance, geostationary satellites are extremely faint. They are well beyond the limit of visibility of either binoculars or small telescopes and, indeed, require sizeable instruments with professional-style equipment to detect them. There may be instances, however, when a satellite's solar panels or other surfaces reflect sunlight exactly in the direction of the observer, making them bright enough to be seen with binoculars. Such chance alignments are rare.

Another listener asked if geostationary satellites would twinkle like stars if you could see them, and the answer is that yes, they would. They are so far away that they are effectively point sources of light, so their light rays are distorted by the Earth's atmosphere in exactly the same way as starlight (see 'Twinkle, Twinkle...' in the previous chapter). The only thing that would distinguish a geostationary satellite from a star is that as the Earth rotates, a star appears to drift from East to West. The satellite, being geostationary, doesn't.

I saw a brilliant, star-like object appear for a couple of seconds – what was it?

It was almost certainly an Iridium flare. And while you might be forgiven for assuming that this is some super-exotic natural phenomenon deep in the Universe, I'm afraid you'd be mistaken. It's a satellite.

Iridium is an example of an alternative approach to global communications. Instead of having a few satellites in geostationary orbits, you could have lots of satellites in low-Earth orbit. But why would you want to do that, given that the greater number of launches would inevitably make it a more expensive proposition? The answer lies in the particular application. Iridium is a satellite phone system, involving direct up-and-down links to mobile phones and pagers. A hand-held phone has nothing like the transmission power of a fixed ground station, and can therefore only communicate over relatively short distances. Moreover, shorter distances mean a smaller time-delay than that of a geostationary satellite, for which the out-and-back signal time is about a quarter of a second.

Iridium consists of no less than 66 satellites in polar orbits (i.e., orbits that pass over the Earth's poles) at a height of 780 kilometres. They are spaced around the Earth in six orbital planes, each of which contains 11 operational satellites chasing one another around the planet. That way, complete global coverage is provided.

An Iridium satellite is barely visible to the unaided eye under normal circumstances, but at any given point on the Earth's surface it will occasionally flare to become the brightest object in the night sky after the Moon, easily outshining the dazzling planet Venus. Such a flare will last for a few seconds, after which the satellite fades again to near-invisibility. What is happening is that one of the satellite's three, flat-panel antennas is catching the light of the Sun, and beaming it down to the observer. The panels are highly reflective and each one acts just like a large, space-borne bathroom mirror.

As the satellite travels along the sunlit part of its orbit, therefore, a moving spot of light some 10–20 kilometres across is thrown onto the ground, and any observer along its path will witness a flare. This is a very dramatic event when seen unexpectedly, as the listener who asked this question had discovered. However, forecasting when you might see a flare is as easy as visiting the Heavens Above website (www.heavens-above.com/), and entering the name of your city, town or village. The Iridium flare predictions given by this service are extremely accurate, and you will be surprised at how frequent they are.

Why do you sometimes see multiple satellites, often travelling North–South?

While it is quite common to see chance pairings of otherwise-unrelated satellites crossing the sky at the same time and in similar paths, there are also spacecraft that orbit in formation for various reasons. Sometimes, for example, the International Space Station can be seen, with the Space Shuttle orbiting a few hundred kilometres ahead of or behind it during a servicing mission. Both these objects are unmistakeably bright, and this is quite a spectacular event.

Other missions require satellites to be paired permanently in orbit. A well-known example of this is the attractively named GRACE project (Gravity Recovery And Climate Experiment), which senses variations in the Earth's gravity by flying two satellites in the same orbit, one some 220 kilometres ahead of the other. The trick of sensing gravity relies on the separation of the two satellites being known rather accurately – to a mind-blowing 0.01 *millimetres*.

Most multiple satellites, however, are military in origin, so public information about them is very limited. Like GRACE, to maximise their coverage of the planet's surface, they usually are in polar orbits and are hence seen moving across the sky in a North–South or South–North direction. The kind of sensing carried out by these reconnaissance satellites might involve one craft carrying a radar transmitter and its companion(s) the receivers, hence the need for formation flying. Best not to mess with them.

Do you get a good view of the International Space Station with the Anglo-Australian Telescope?

Since the Anglo-Australian Telescope (AAT) is Australia's largest optical (visible light) telescope, you might expect that it would give you a great view of absolutely everything in the sky. Unfortunately, though, like most large telescopes, the AAT is entirely unsuited to observing spacecraft and, to the best of my knowledge, has never done so.

The reason for this is the way the AAT is pointed at celestial targets. Its mounting was designed to compensate for the leisurely rotation of the Earth in tracking objects across the sky. When you are observing a satellite, however, the dominant motion is the satellite's own, which often amounts to tens of degrees per minute in angular speed. This is far more than the AAT can cope with.

There are other, more specialised instruments for imaging spacecraft (called, not surprisingly, satellite tracking cameras) which are mounted in such a way that under computer control they can be swung to follow the rapid motion of the target. Their effectiveness is demonstrated by events back in 1981, when a military camera was used to look for any gaps in the heat-shielding tiles of the first Space Shuttle, *Columbia*, on her maiden orbital flight. Although some of the images were shown briefly on TV at the time, they have never formally been made public.

Images of the International Space Station have also been made with similar equipment, of course. However, one of the nicest ground-based portraits taken recently was by a French amateur astronomer, Thierry Legaut, who caught the Space Station – accompanied by a visiting Space Shuttle, *Atlantis* – in silhouette as they crossed the disc of the Sun. This was with a telescope designed for solar observing and, not surprisingly, a key ingredient was split-second timing.

For the record (and in answer to another listener's question), the International Space Station is in a low-Earth orbit at a height of some 340 kilometres above the surface. Because it is very large (its solar panels alone have an area of a third of a hectare), the drag of the upper atmosphere causes its orbit to decay rapidly and it requires frequent rocket burns to maintain height.

What is the minimum distance from Earth you would have to be to see the complete disc of the planet?

This is a question with more than one possible answer, depending on how you interpret it. If you mean what distance do you have to be to see a complete hemisphere, then the answer is infinity, because the nearer you get to a sphere, the less you can see of its surface. In reality, however, you can see most of a hemisphere of the Earth from the distance of the Moon, except for a band about 100 kilometres wide around the edge of the disc. As you approach the Earth, that unseen area grows larger, so the disc you can see includes a progressively smaller proportion of the Earth's surface.

When I queried the listener who raised this question, however, it turned out that this was not what he meant. He wanted to know how far away you'd have to be to see the Earth as a disc. That is, a large, circular object stuck in front of your eyes. Even this doesn't have a straight answer, though, because you have to specify the size of the disc you want to see. To cut a long story short, let's suppose you decide that the Earth will look like a disc if it makes an angle of 90 degrees at your viewpoint. In such a case it will fill a large fraction of your field of vision, but you will still be able to see darkness all the way around its edge.

It turns out that the height you would need to be above the surface to view this spectacle is 2,640 kilometres. And that's a rather dangerous place to be, since it's in the thick of the Earth's radiation belts. While you're there, you'll be able to see an area of the surface some 10,000 kilometres across

– a quarter of the Earth's circumference. Best not to linger, however...

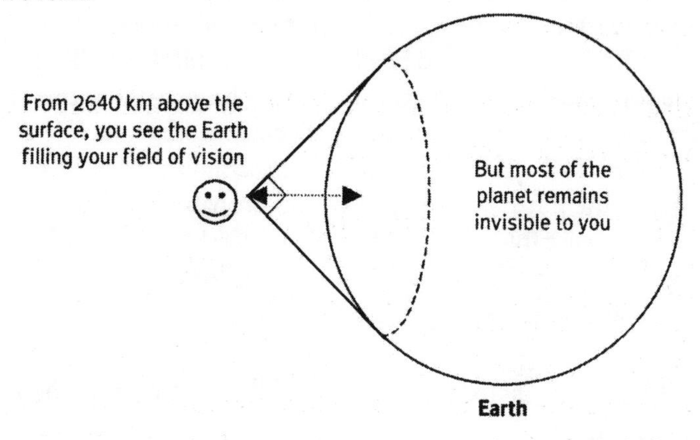

You need to be 2640 km above the Earth for the disc of the planet to fill your field of vision by making an angle of 90 degrees at your eye. Watch out for the radiation belts.

EXOTICA ASTRONAUTICA: SOME SPACED-OUT IDEAS

Could you dump nuclear waste into the Sun?

This is a great suggestion from an admirably pollution-conscious listener, but unfortunately it's so fraught with difficulties as to be unworkable – at least for now. The world's nuclear reactors produce some 15,000 tonnes of unprocessed high-level waste annually, so the first problem is the sheer size of the payload. Then there is the decidedly tricky political issue of launching said waste through the Earth's atmosphere aboard a rocket that, with the best will in the world, is still a risky piece of kit. The handful of deep-space probes equipped with nuclear power generators containing just a few kilograms of radioactive plutonium (such as *New Horizons*, currently *en route* to Pluto) have always caused controversy. Imagine how the launch of *tonnes* of waste might be greeted...

Finally, there is the daunting problem of dropping it into the Sun. To do this, it has to be placed in an orbit that will intersect the solar surface, and that will require a huge amount of energy. Effectively, the nuclear-waste spacecraft will have to lose much of the Earth's orbital speed of 30 km/second by firing its rockets against the Earth's motion. It would probably need at least one gravity-assist fly-by of another planet to slow it further. Thus, for all it's a nice idea, it's far from an optimum solution. Much easier and safer to process the waste and bury it.

Are there any spacecraft leaving the Solar System, and where are they going?

There's something rather special about spacecraft that have been given enough of a kick to escape the Sun's gravity altogether. These are human artefacts that will journey through space for a *very* long time, and may even outlast humankind – and, indeed, the Earth – altogether. If that's not a sobering thought, I don't know what is.

At the time of writing, there are five of them. Their names are *Pioneer 10* and *11*, *Voyager 1* and *2* and – the most recent member of this exclusive club – *New Horizons*. The *Pioneers* were launched in 1972 and 1973, the two *Voyagers* in 1977, and *New Horizons* in January 2006. The four older spacecraft had their moments of glory in the 1970s and 1980s when they gave us our first close look at the giant planets of the outer Solar System, but *New Horizons* is yet to shine. It will fly past the dwarf planet Pluto and its moons Charon, Nix and Hydra in 2015, and will then go on to rendezvous with an as-yet unspecified object in the Kuiper Belt, that remote disc of icy debris left over from the birth of the Solar System. After that, it will make its way ever outwards into the depths of space.

New Horizons is, in fact, the fastest of these spacecraft, with a current speed away from the Sun of 23 km/second. At this rate, if it was pointing in the right direction, it could expect to reach the nearest star in some 60,000 years. But don't bother to wait around, as not only will it slow down appreciably as it escapes the Sun's gravity, but it's not actually going that way anyway.

More interesting in terms of what they can tell us about the Sun's environment are the two *Voyager* spacecraft. At a current distance of 15 billion kilometres from the Earth (about three times further away than Pluto), *Voyager 1* is the most remote human-made object in existence, while *Voyager 2* is some three billion kilometres behind it. Although both spacecraft are on the same side of the Solar System, *Voyager 1* is climbing above the plane of the Earth's orbit and *Voyager 2* is dropping beneath it. They are both still active, and currently sending back information about the Sun's magnetic sphere of influence.

Both have now reached a region with the startling name of the termination shock, which is where subatomic particles from the Sun collide with those in interstellar space (that is, in the Sun's wider environment). Together with an outer, all-enveloping 'heliopause', the termination shock defines the bubble within which the Sun is the dominant influence, and it seems from the *Voyager* measurements that this bubble is slightly dented in the south. That is possibly due to the influence of the underlying interstellar magnetic field.

One further little mystery associated with these escaping spacecraft is the so-called *Pioneer* anomaly. When we were navigating through space a few questions ago, we noted that the position of a spacecraft is very predictable, once its initial position and velocity are known. How surprising, then, that neither of the *Pioneer* probes was quite where it should have been when contact with them was lost a few years ago. In *Pioneer 10*'s case, the error amounted to about 400,000 kilometres – the distance between the Earth and the Moon.

Some scientists have suggested that this discrepancy reveals a flaw in our understanding of gravity, and that we should be rewriting the textbooks. Much more likely, however, is that some non-gravitational phenomenon such as the solar wind, a temperature effect or, simply, a fuel leak, has not been allowed for properly. The jury is still out on this one, but no doubt an answer will eventually be found.

The ultimate fate of all these spacecraft is almost certainly to fall within the gravitational influence of another celestial object – most likely a star, or one of its planets or their moons. The probe will either collide with the object, or will be captured into orbit around it. A less likely, but not impossible fate, is that one of them will be picked up by representatives of an extraterrestrial civilisation. It is for this reason that both the *Pioneer* and *Voyager* spacecraft carry little souvenirs of humanity.

The two *Pioneers* carry a gold plaque giving directions to Earth, and showing the aliens how nice and tasty we look by depicting naked human figures. The *Voyagers*, however, tote a 12-inch long-playing record (remember them?) of the sounds and sights of the 1970s, complete with a stylus and cartridge to play it with. That should be enough to put any hungry aliens off.

What is the space elevator and will it work?

One of the basic problems of contemporary spaceflight is that the only way of getting stuff into orbit is by taking it on board a large, expensive, risky, polluting rocket. That puts the price up – to something like A$30,000 per kilogram. How much more would we be able to exploit space if the cost came down to a few dollars per kilogram? That's the thinking behind the space elevator, one of the truly audacious ideas of the Space Age and one that has received increasing attention during the past few years.

The idea of a tower into space goes back to the end of the nineteenth century but, once again, it was Arthur C. Clarke who painted our modern picture of the elevator, in his 1978 novel, *The Fountains of Paradise*. Imagine one of Arthur's geostationary satellites, discussed earlier in the chapter. It remains fixed in position over the Earth's equator at a height of 35,786 kilometres. Now let's suppose you could lower a cable from the satellite towards the Earth, at the same time counterbalancing it with another cable extending outwards in the opposite direction. That way, the satellite's centre of mass would remain at the geostationary point. So far, so good.

Now, you keep on lowering and extending, lowering and extending, until, *voilà*, the bottom of your cable meets the top of a rigid tower 50 kilometres high, built on the Earth's equator directly underneath the satellite. Oh, didn't I mention the tower? Anyway, you quickly tie the cable to the tower, whereupon it becomes a tether rather than just a cable, and you have the basics of a space elevator. Then, if you can strengthen the tether, and equip it with the right kind of rails or guides, you can run vehicles up to the geostationary point and back for, well, just a few dollars per kilogram.

Once the system is in place, it opens up all kinds of possibilities. A full-scale space station could be built at the geostationary terminal, and this could be used as an engineering base to assemble very large spacecraft for further exploration of the Solar System, the components being ferried up from Earth on the elevator. The fuel required to propel such craft would be much less than that needed for a conventional mission lifting off from the Earth's surface.

One model of the system envisages relatively leisurely elevator speeds of around 200 km/h, which would require a week for the elevator to reach the top. That raises problems for human travellers, however, because they would be in the

Earth's dangerous radiation belts for several days. Higher speeds may be essential.

The space elevator has been the subject of a number of serious research investigations, including recent studies at the Advanced Projects Office of NASA's Marshall Spaceflight Center. Promising technologies for the tether's structure are emerging, such as high-strength carbon nanotubes. Tests with high-flying tethered balloons have already been carried out. And, to shorten the balancing cable, it has been proposed to use a massive natural object as a counterweight – to wit, a small asteroid. That kind of engineering is certainly beyond present-day technology, but within half a century, it's likely that we will have the necessary know-how at our disposal. Then, it's just a question of saving up the money (no one is guessing how much), and building it.

Will it work? Probably. But the timescale is the big uncertainty. When Arthur C. Clarke was asked when we might see a space elevator completed, he is reputed to have answered, 'About 50 years after everyone stops laughing.' It seems to me that the space industry is just getting over its last few chuckles.

Could we terraform Mars?

If the space elevator is audacious, this idea is truly barmy. It relates to the possibility of providing an 'overflow' planet for Earth's burgeoning population, a place that we can colonise in the distant future. And of all the planets in the Solar System, Mars is the one most like Earth. Although its gravitational attraction is only one-third that of the Earth, the planet's day (24h 39m 35s) and axial tilt (25 degrees) are both very similar to ours. On the other hand, the average surface temperature is around –63°C, and the atmospheric pressure is less than 1 per cent of Earth's. The atmosphere is also extremely dry, although all recent evidence suggests that the Martian soil is rich in water ice.

The idea, then, is to modify Mars' atmosphere to make it more like the Earth's, and thereby 'terraform' the planet. On a global scale, this would be a monumental undertaking and could take centuries, perhaps even millennia. Possible mechanisms do exist, however. They depend on the production of vast quantities of greenhouse gases to thicken and warm the planet's atmosphere. The greenhouse gases could be produced artificially, or by warming frozen carbon dioxide at the planet's poles. A rather more extreme suggestion involves taking aim at Mars with an ammonia-rich asteroid from the Sun's distant Kuiper Belt to warm things up a bit, and seed the atmosphere with greenhouse gases.

More modest ambitions include using kilometre-sized orbiting mirrors to reflect sunlight onto small areas of the Martian surface a kilometre or so across. If the local surface temperature in these pockets could be increased to around 20°C, astronauts working on the surface would not require heavily insulated clothing and accommodation, as they would at present. Moreover, ground water might be obtained by local thawing. Even this is ambitious, however, as the deployment of such space mirrors would be far from straightforward.

The idea of terraforming Mars also raises some complex ethical questions. Do we *really* want to trash another planet as we have our own? More prosaically, however, it's possible that the need may never arise. Long before we could make Mars habitable, we surely must have learned how to control our own rapacious demands on the Earth's resources. Or have perished.

AND FINALLY...

Who owns space junk?

When I was asked this question by a listener, I had an uncomfortable feeling I knew precious little about space law and, as it turned out, I was right. At least I admitted it, because the truth is that very few people understand space law – even the lawyers.

Space law is primarily embodied in the UN-ratified Outer Space Treaty of 1967 and its four additional conventions of 1968–79. The problem is that these regulations were formulated in the Cold War era, when the principal users of space were a handful of superpowers. Today, the exploitation of space has extended not only to the commercial sector but also to the tourist industry, and the present laws are totally inadequate to deal with the possibilities that might arise. Two commonly cited areas of difficulty are the industrial exploitation of space minerals (for example, from asteroids), and the control of intellectual property rights in collaborative ventures such as the International Space Station.

The regulation of space debris falls into a similarly grey area, particularly since the term encompasses everything from stray nuts and bolts a few millimetres across to complete satellites and large, spent booster rockets. While the Outer Space Treaty is clear that objects in space remain the property of their original owners, the mechanism for claims against any damage caused by these objects is far from straightforward (see www.unoosa.org/oosa/SpaceLaw/outerspt.html). Because of this, the European Space Agency's Director of External Relations, René Oosterlinck, recently stated (in an ESA podcast of 9 November 2006) in relation to space debris that:

> There are no existing laws to cover this type of pollution at the moment. In these and many other fields, the Outer Space Treaty no longer matches reality, and the time is ripe for review.

Which just about sums it up.

In a final twist to this story, an Australian archaeologist, Alice Gorman of Flinders University, has recently suggested that certain items of space junk should be heritage-listed as important artefacts of human history. She wants to see this happen in advance of any 'clean-up' that new laws on space debris might precipitate. Perhaps the number-one candidate for such a listing is the oldest human-made object in space, which brings us neatly back to where this chapter started. For that object is the US Navy's belatedly successful *Vanguard 1*, launched on 17 March 1958 – the second US space vehicle to enter orbit. Unlike some items of space debris, whose orbits decay rapidly, *Vanguard 1* is expected to have a lifetime of some 240 years. So there's still time to think about it...

CHAPTER 6

GREEN CHEESE NO LONGER

EARTH'S ESSENTIAL COMPANION

It's the most visible object in the night sky. Yet sometimes it's not there. It is the only other world on which humans have walked. Yet a vocal minority of the population apparently believes that never happened. It has played a major role in the evolution of life on Earth. Yet some scientists curse it for spoiling the darkness of their skies. Everyone knows about it, and – love it or hate it – we're stuck with it. For the time being, at least.

The Moon is the best-known of all astronomical objects. Even toddlers know what its crescent looks like – usually before they have learned how to draw stars. And perhaps such consummate familiarity breeds a degree of contempt for this celestial body that, in reality, ranks with the Sun and the Earth itself in its importance to our existence and wellbeing.

Most people know that the Moon is an airless world with a diameter rather more than a quarter of the Earth's (3476 kilometres) and with one-sixth of the Earth's gravity. Its distance from us – 384,000 kilometres – is only 30 times the Earth's diameter. But getting there would be a five-month drive at the motorway speed limit (without any rest stops), bringing a whole new dimension to 'Are we there, yet?'

Because the Moon's diameter is a largish fraction of the Earth's, the pair were regarded for many years as a 'double planet' rather than a planet plus its satellite, with the two components perhaps having had a common origin. We now believe that the Moon formed after the Earth – though still very early in the 4.6 billion-year history of the Solar System – when an object the size of Mars dealt the Earth a glancing blow. The end product of this collision was that debris from the Earth's outer layer (its rocky mantle) collected in orbit around the planet, gradually coming together in a process known as 'accretion' to make the Moon. This explains the Moon's differing geology from the Earth's, as revealed by the 380 kilograms of rock and soil samples returned by the *Apollo* astronauts.

To the unaided eye, the Moon appears to be made up of greyish areas still known as *maria* ('seas'), and brighter highland regions. While the belief in *maria* as ancient oceans has long been disproved, we do know that they represent low-lying volcanic plains – the result of lava flows. So yes, they were indeed once wet, but rather hotter than your average ocean. When viewed through binoculars or a small telescope, the highland regions show both craters and mountains, in stark contrast to the smoother *maria*. Both the craters and the mountains are geologically much older, rising higher than the lava flows, and the craters record the scarring created by asteroid bombardment during the Solar System's infancy. This cratering record tells us much about conditions during that early period.

The Moon, of course, is a subject of great interest to radio listeners. Rather surprisingly, the majority of the questions concern the Moon's general appearance, and its comings and goings in the sky, rather than its physical details. This is reflected in the selection below. A few listeners have asked really intriguing questions, however. What would it be like on Earth if the Moon wasn't there? Is the Moon drifting away from us? And then there is the remarkable issue of why, in an almost mystical coincidence, the Sun and Moon should appear to be the same size in the sky, allowing perfect solar eclipses to take place. Read on, folks...

SHEER LUNACY: THE MOON'S PHASES

Why do you sometimes see the Moon during the day?

While everyone learns about the phases of the Moon in school, it's a lesson easily forgotten. If you're not thinking about this sort of thing all the time – and most of us aren't – then it's easy for the whole deal to become shrouded in mystery, and for little puzzles to emerge about the Moon's appearance and shape. Why don't we always see it at night? Why can we sometimes see it during the day? Oh, and what was all that stuff about waxing and waning again...?

These – and a few other questions – have all been asked by listeners, and I'd like to lump them together in a short guided tour of the Moon's phases. The best way to take such a tour is with a tennis ball and a marble, one in each hand. Set up a desk lamp some distance away to represent the Sun, and then make the marble (the Moon) go around the tennis ball (Earth) in an impression of the Moon's orbit, placing your eye near the tennis ball to watch the changing illumination of the marble. If you're under eight, you can make motor-bike noises to accompany the process, but that's optional. The almost inevitable outcome is that you will drop both the ball and the marble, and everyone will laugh at you as you chase them around the floor in embarrassment.

But the point has been made. The Moon revolves around the Earth (anti-clockwise as seen from above the North Pole) in a period of 27 days 7 hours 43 minutes. Remember, though, that the Earth is also moving around the Sun (again in an anti-clockwise direction), so the time between successive occasions when the Sun and Moon lie in the same direction is a bit longer, because the Moon has had to travel slightly further than one complete orbit. This period is 29 days 12 hours 44 minutes, and is technically known as a synodic month. From it is derived its human-made counterpart, the calendar month – which has evolved to become the twelve roughly equal chunks into which the year is now divided.

The Moon lies in the same direction as the Sun at the time of a 'new Moon', and this represents the start of the lunar month, or lunation. It doesn't automatically result in an eclipse of the Sun, because the Moon's orbit is tilted at 5 degrees to the Earth's. Usually, therefore, the Moon steals above or below the Sun in the sky. Either way you don't see it, because its dark face is presented to us, while the Earth's atmosphere is brightly illuminated by the Sun.

As the Moon then progresses around its orbit, we see first a slender crescent setting shortly after the Sun in the evening sky. From night to night, the Moon sets progressively later as its crescent broadens until, seven or eight nights into the lunation, it takes on the semicircular appearance we call 'first quarter'. The angle between the Sun and the Moon is then 90 degrees, and the Moon is high in the sky at sunset.

During the following week, the Moon continues to grow beyond a semicircle, becoming 'gibbous' – a curious word derived from the Latin for 'humpbacked'. It eventually becomes a perfectly circular full Moon on the 14th or 15th night. Our satellite is then opposite the Sun in the sky, rising at sunset and setting at sunrise, but again, is only darkened by the Earth's shadow in a lunar eclipse if the alignment is right. The Moon then shrinks gradually towards 'last quarter', after which it becomes a crescent again, now rising before the Sun in the pre-dawn sky.

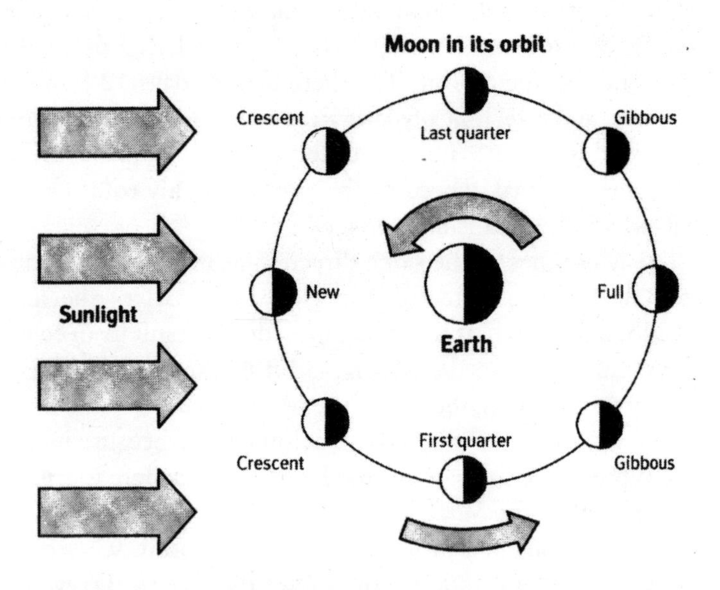

Phases of the Moon as conventionally shown from above the Earth's North Pole

Phases of the Moon—the usual boring textbook picture.

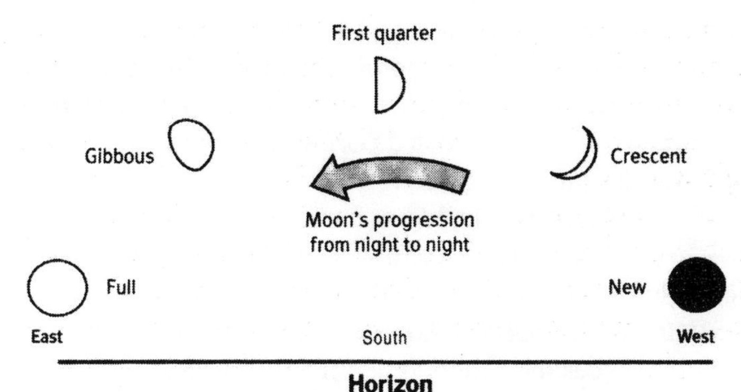

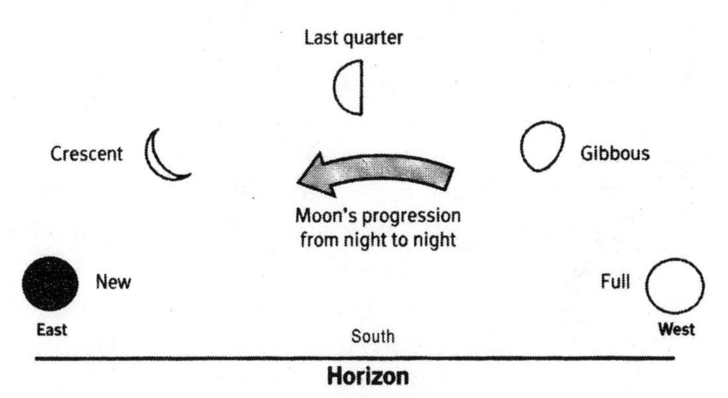

How we see the phases of the Moon from the northern hemisphere.

It's when the Moon is in its gibbous phase before and after full Moon that it is most likely to be seen during the day. Even though a crescent Moon is actually in the daytime sky for longer (being closer to the Sun), it is not as conspicuous, so is often missed.

Another listener asked how you can tell whether the Moon is waxing (growing after new Moon) or waning (shrinking before new Moon), and that's easy, too. For observers in the northern hemisphere, if the illuminated portion of the Moon is on the right (towards the West), then it's waxing. Conversely, if the illuminated portion of the Moon is to the left (East), it's waning. These rules are reversed in the southern hemisphere, where the waxing Moon is illuminated on the left (which, perversely, is still its western portion, however), and vice versa.

See also: Does the Moon look different when seen from the northern and southern hemispheres?

Why does the direction of moonrise or moonset change from night to night?

We might as well get all the geometrical stuff out of the way at once. At first sight, the progression of moonrise and moonset backwards and forwards along the horizon appears baffling, but in fact it's closely related to the way the Sun behaves (see 'Why does the direction of sunrise or sunset change throughout the year?' in Chapter 3). Since the Moon's orbit is tilted at only 5 degrees to that of the Earth, the Sun and Moon follow very similar paths through the sky. This means that the discussion on the changing direction of sunrise and sunset holds broadly true for the Moon.

The big difference, however, is that compared with the Sun, the Moon positively whizzes through the heavens. It goes all the way around the sky in a month, or 'Moonth' (of 29.53 days), whereas the Sun takes a whole year to get around. Consequently, the Moon appears to move very quickly among the stars. Very roughly, from one night to the next, its position will change by 10 to 15 degrees (rather more than the width of your fist at arm's length), always progressing towards the East.

The result of this brisk behaviour is that the nightly change in the direction of moonrise and moonset is dramatically faster than the progression of the Sun's rising and setting. It relates not only to the time of year, but also to the time of the month (or, in other words, the phase of the Moon). When the Moon is full, it rises opposite the Sun in the sky. A fortnight later, when it's new, it rises in more or less the same direction as the Sun. This difference in the direction of moonrise (or moonset) between the new and full phases is most pronounced near the solstices, and least near the equinoxes.

Does the Moon look different when seen from the northern and southern hemispheres?

Yes, it does, because what we think of as 'up' is different in the two hemispheres. The best way to visualise this is to imagine yourself standing at the North Pole, looking horizontally towards the Moon. If you then travelled to the South Pole, the Moon would obviously be upside-down compared with its appearance before.

That doesn't just apply to the Moon's physical features, but also to its phases. We saw a couple of questions ago that a waxing (growing) Moon is illuminated on the right in the northern hemisphere, but on the left in the south. (As you cross the equator, of course, the Moon passes more or less overhead, so you can choose which way up you want to look at it.)

I once fell foul of this effect in an embarrassing way, and I'd be most grateful if you would keep this story to yourself. Almost thirty years ago, when I was an astronomer at the Royal Observatory in Edinburgh, my work brought me to Australia for the first time, and I will never forget my first glimpse of the southern sky. It was one of the worst moments of my life. It was through the window of a Boeing 747, on the last leg of its long journey down to Sydney. Gazing out into the pre-dawn sky, hoping to catch a glimpse of some of those fabled southern constellations, I saw the Moon. It was at first quarter. In seven days' time, it would be full, and flooding the night sky with so much light that the faint stars I had come all this way to observe would be absolutely invisible. Oh, no! It was a blunder of monumental proportions – getting the phase of the Moon wrong for an observing run on Australia's biggest telescope. Would I ever be able to look my colleagues in the eye again? My career was over...

It took me fully ten minutes to convince myself that what looks like first quarter in the northern hemisphere is actually last quarter in the south. As we've just seen, the half that is illuminated by the Sun is not the half you'd expect to see lit up if you're used to the northern sky. So, in seven days' time, the Moon would actually be new, not full. The sky would be dark, save for the glorious panorama of the southern stars, and those faint objects would begin yielding up their secrets. What a relief. In the event, the observing run was completely washed out by bad weather – but at least I now had my Moon phases right.

MOONSHINE: EARTH'S GOOD-LOOKING SATELLITE

Why does the Moon look transparent when you see it in daylight?

This was an impression of the Moon that had never occurred to me until a delightful lady raised it on air some years ago. And she was right – it does sometimes look transparent, when the blue of the daylight sky seems to shine through the gossamer outline of the Moon. In fact, it's the other way around. As we saw in Chapter 4 ('Can you see stars in daylight?'), the daytime sky is itself transparent, whereas the Moon is an object in the distant background being seen through our luminous atmosphere. Thus, the blue of the sky in front of the Moon combines with its own pale light to give it a bluish tinge – and hence the translucent effect that the listener asked about.

Why does the Moon look so big when it's low in the sky?

Of all the phenomena relating to the Moon, this one is undoubtedly the most frequently remarked upon. It is especially obvious around the time of full Moon, when the disc of our satellite looks positively *huge* as it rises at dusk in the eastern half of the sky. And the full Moon appears similarly enormous when it sets towards the West as dawn breaks. The effect is extremely convincing – psychologists tell us that the Moon seems to be two to three times larger near the horizon than when it is high in the sky.

You'll have noticed I said 'psychologists' there, and not 'physicists'. That is because this effect is not real, but takes place entirely inside our own heads. It is the human brain that causes the apparent increase in size, not the physics of the atmosphere or the geometry of the Earth–Moon system. The effect is known famously as the 'Moon illusion'.

There are many simple ways of proving that it is, indeed, a psychological illusion. A thin pencil held at arm's length, for example, will just about cover the Moon whether it's high in the sky or low on the horizon. And photographs of the Moon taken in the two situations prove the same thing.

In fact, a little thought shows that when the Moon is low on the horizon, it is actually slightly further away from the observer than when it is high in the sky, so it should look *smaller* rather than larger. Imagine that the Moon is exactly overhead. This brings it nearer to the observer than when it is seen horizontally by a distance equal to the radius of the Earth, some 6,400 kilometres. That amounts to a difference in distance (and hence in the apparent size of the Moon) of almost 2 per cent – a significant change in astronomical terms, although not quite perceptible to the unaided human eye.

Physics produces another effect that makes the Moon look a fraction smaller – at least in one dimension – when it is very low in the sky. Rays of moonlight from a near-horizontal Moon strike the upper layers of the Earth's atmosphere at a much shallower angle than when the Moon is high up, and the result is a slight downward bending of the light rays. The phenomenon is called 'atmospheric refraction', and the end product for the observer is that the Moon appears to be a little higher in the sky than it really is. More significant, though, is the fact that rays from the lower edge of the Moon are bent a tad more than those from the upper edge, so that the Moon's disc appears slightly squashed in the vertical direction. Not as much as a rugby football, but enough to be noticeable – especially through binoculars.

So if the physical reality is that a low-down Moon appears slightly smaller than a high-up one, why do our eyes tell us it looks so big? This conundrum of perception has long been a subject of debate among cognitive psychologists. There is at least one hefty volume of learned text on the topic, written during the late 1980s by some of the world's foremost experts. Browsing through this tome gives the distinct impression that psychologists were (and perhaps still are) groping to understand why our minds tell us so convincingly that the Moon is bigger than usual when it's simply not true.

It used to be thought that the Moon illusion happens because a low Moon is surrounded by familiar objects such as hills, trees and buildings, while a high one is seen in isolation. Thus the horizontal Moon is seen at its 'real' size, but the lonely Moon high in the sky appears diminished. There is nothing familiar to compare it with, so it shrinks. More recent thinking suggests there's more to it than this, however, and the most popular of today's theories relates our impression of the size of the Moon to the perceived shape of the sky itself.

It has been known for many years that we think of the sky not as a hemispheric dome over our heads, but as a flattened dish. In other words, our perception is that the sky is much further away near the horizon than it is over our heads. It's no mystery why this should be so – those fluffy clouds we often see scattered uniformly over a fair-weather sky, for example, appear to become smaller as our gaze moves down from the point overhead (the zenith) towards the horizon. Of course, in reality, they're all similarly sized, but the effect of perspective makes them seem to shrink. On a very clear day in flat country, where the view is uncluttered by trees and buildings, clouds near the horizon can appear perhaps 50 times smaller than their local counterparts because they are so much further away. No wonder our brains tell us that the sky is more like an upside-down dinner plate than an upside-down fruit bowl.

Now, into this trick of perception steps the Moon. We most often see it high up, so we have a built-in impression of how big it should look. As it rises over the horizon it is, as we have seen, essentially the same size as it always is. But hang on a minute – the human mind says that anything near the horizon is a long way off, and therefore must look small. The Moon, however, is the same size, challenging the brain to make sense of what is going on. How does the brain respond? There are no prizes for guessing that it tells us the Moon is bigger than it should be, because it doesn't look as small as the brain thinks it should. Hence the Moon illusion.

It's not just the Moon that undergoes this illusory swelling near the horizon. The same thing happens to the Sun, although this is usually more difficult to spot because of its brilliance. But have a (careful) look next time the Sun is low in the sky and dimmed by haze or fog, and you'll see that it does look bigger than you'd expect. More subtly, familiar star patterns also look huge near the horizon. Most notable is the constellation of Orion, which rises and sets exactly on the East and West points of the compass almost everywhere in the world. Unfortunately, many places lack the ingredients needed to observe this phenomenon – dark, pollution-free skies all the way down to a low, uncluttered horizon – but when it is seen it is very striking. The Heavenly Hunter looks absolutely enormous.

Finally, many commentators have noted that the Moon illusion can be made to disappear by standing with your feet astride and bending down to look at the Moon between your legs. Personally, I have always thought this was a cunning ploy on the part of astronomers to make the rest of us look daft, but there is probably something to be said for radically changing the way you observe the scene. A bit more speculatively (since there's clearly no need to adjust your nether garments to observe the Moon), I wonder if this could be where the expression 'mooning' comes from...?

Why does the Moon sometimes appear orange or red?

This effect often goes hand-in-hand with the one in the previous question, because the orange colour is usually noticed when the Moon is low on the horizon. However, in this case, the Moon's unusual appearance is due to a physical rather than a psychological effect. The orange colouring of the Moon is exactly analogous to the reddening of the Sun mentioned in Chapter 4 ('Why is the sky blue?'), and is caused by the same mechanism. Near-horizontal moonlight passing through a thick layer of the Earth's atmosphere is scattered preferentially in the blue region of the rainbow spectrum of colours, leaving a mixture of hues biased towards red. The end product is a yellow or orange Moon – and it is often quite striking. In contrast, when the Moon is high in the sky, it looks white.

An occasional exception to this is when the full Moon passes through the shadow of the Earth during a lunar eclipse. If the eclipse is total (that is, the Moon passes through the central region of the shadow), then no direct sunlight falls on the Moon's surface. It is still illuminated, however, by sunlight filtering through the Earth's atmosphere, light that has had its blue component removed by scattering. The surreal result is that from the Earth, the Moon appears dimly illuminated with a coppery red glow. Truly, the Moon is turned to blood. If you were standing on the Moon's surface during the eclipse, however, you would see something even more spectacular – the black disc of the Earth, rimmed with a brilliant red atmosphere.

Why does the illuminated part of the crescent Moon seem to point above the Sun?

This is an effect best observed in the afternoon sky when the Moon's phase is a day or two before first quarter. The Moon then has the shape of a broad crescent, but is sufficiently far from the Sun for the phenomenon to be really noticeable. I'm afraid I have to resort to rather poetic metaphors to describe it, so I hope you'll bear with me...

If you imagine the crescent Moon as a bow shooting an arrow, then the arrow will point towards the Sun – naturally, since the 'bow' is simply the side of the Moon illuminated by the Sun. But when the Moon is a reasonable distance from the Sun the 'arrow' points in a direction which, if extended across the sky in a straight line, would pass above the Sun. The reason for this is to do with perspective. Light rays from the Sun travel in straight lines, of course, but when they go from one side of the sky to the other, they appear to arch upwards rather than simply travelling parallel to the horizon. In fact, they form great circles on the spherical surface of the sky (which, by the way, is usually known as the celestial sphere, because it also includes that part of the sky that lies beneath the horizon).

A great circle is the shortest distance between two points on a sphere and, on a globe of the Earth, for example, it appears to be a straight line when viewed from the correct angle. But it curves with respect to the equator. The situation is the same with the sky. When you follow the direction of the 'arrow' from the 'bow' of the crescent Moon across to the Sun, you have to imagine it following a curved path through the sky rather than a straight line parallel to the horizon. Then you'll see it's pointing in the right direction.

I'm sorry to have to tell you that no one to whom I've given this explanation has ever believed it, so I'm clearly not explaining it very well. It is, however, the correct one... Trust me.

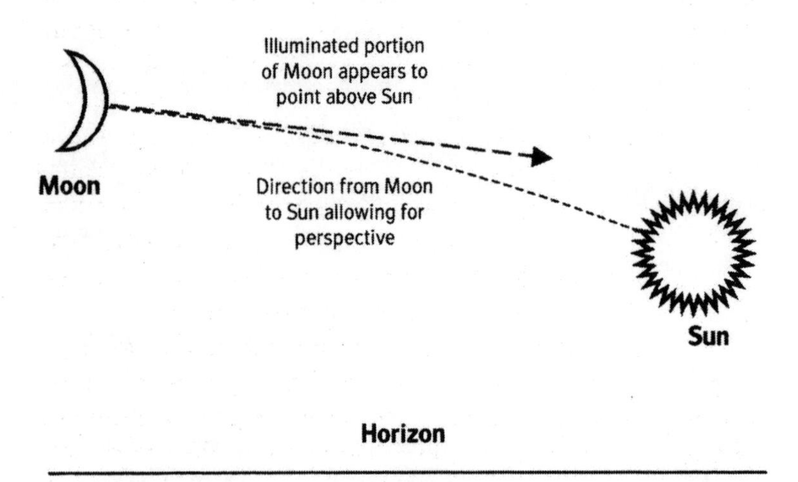

Rays of sunlight illuminating the Moon travel in straight lines, but perspective fools us into thinking they curve across the sky.

What is earthshine?

Leonardo da Vinci correctly explains the origin of the beautiful phenomenon known quaintly as 'the old Moon in the arms of the new' in his notebook now called the Codex Leicester, written in about 1509. The faintly illuminated lunar disc is visible between the horns of the crescent Moon, and this is most commonly seen for a few days after new Moon in the early evening sky. It's also visible for a few days before new Moon, but as the thin crescent is then rising in the pre-dawn sky it is seen by fewer people.

Once called the 'ashen light' (a term now reserved for a similar phenomenon of unknown origin on the planet Venus), the effect is today known as 'earthshine' – since that's exactly what it is. If you imagine yourself standing somewhere on the dark portion of a crescent Moon, the brightest object in your sky will not be the Sun (since that's below your horizon), but the Earth. Moreover, it will be a full Earth, since the disc of our planet shows lunar observers the reverse of the lunar phases that *we* see. New Moon equals full Earth, and so on. The full Earth is an extremely bright object, particularly when there is significant cloud cover over its visible hemisphere. Thus, it lights up the darkened surface of the Moon to a very high degree – much more than the full Moon lights up the terrestrial landscape. As deduced by Leonardo, this illumination is what we see as the old Moon in the arms of the new – or earthshine.

Climatologists and astronomers find earthshine to be a surprisingly useful tool. By monitoring its brightness, scientists studying the effects of climate change can get a continuous estimate of variations in global cloud cover. The technique allows cloud cover to be assessed on a hemispheric basis much more easily than can be done with data from weather satellites, for example. And combining the earthshine data with space-based cloud measurements gives very high-precision estimates of the Earth's capacity to reflect light. These observations have revealed that between 1984 and 2000 the Earth underwent a steady dimming, but it has since brightened up again. How this relates to global warming is still under investigation.

Earthshine also carries more subtle information than just its brightness. Like all kinds of radiation, its rainbow spectrum has coded signals locked into it, and in this case they relate to conditions on the Earth's surface and in its atmosphere. In particular, the light contains the signatures of water, oxygen and other life-sustaining compounds, revealing that the planet is habitable. Moreover, earthshine's infrared spectrum reveals something called the 'vegetation red edge', which is a diagnostic of plant life and its condition – whether it is fresh and green, or dry and withered. These so-called 'biomarkers' allow us to detect the presence of living organisms simply by looking at the averaged light of our planet.

The importance of this is likely to emerge over the next few years, as new telescopes capable of imaging Earth-like planets around distant stars are built. By comparing the rainbow spectra of those as-yet-undiscovered planets with that of Earth, we may get a good idea of their atmospheric conditions, and whether life exists on them. It is even possible that the discovery of industrial pollutants in their atmospheres might reveal the presence of intelligent life. That would truly be the discovery of the millennium.

MOONLIGHTING: THE VIEW FROM OUR SATELLITE

Can you see the Great Wall of China from the Moon?

Sadly, no you can't. In fact, from the distance of the Moon, you can see no human artefacts whatever on the Earth's surface. According to Alan Bean (Lunar Module Pilot on the *Apollo 12* mission in

November 1969): 'The only thing you can see from the Moon is a beautiful sphere, mostly white [clouds], some blue [ocean], patches of yellow [deserts], and every once in a while, some green vegetation'. And, indeed, photographs bear this out.

From low-Earth orbit (such as the 340-kilometre altitude of the International Space Station), many human-made objects including roads, railway lines, cultivated areas and even ships are visible. If you know exactly where to look, you can, indeed, see the Great Wall of China. But the Moon is more than a thousand times further away and at this distance such objects are quite invisible, even with a large telescope.

So where did this mythical idea come from? It seems to have arisen in a book published in 1938 by one Richard Halliburton, *The Second Book of Marvels*, in which he said that the Great Wall would be the only human-made object visible from the Moon. It is a great credit to Halliburton's persuasiveness that 70 years later, his myth is still doing the rounds.

What's the length of a day on the Moon?

Most people know that the Moon always keeps the same face towards the Earth as it glides around its orbit. Of course, its appearance changes with the progression of its phases and the consequent changing illumination of features on the lunar surface, but it's always the same features that we see. The fact that we always see the same face tells us that the length of a day on the Moon is the same as the length of time it takes to orbit the Earth – a synodic month. Remember the synodic month? It's just the interval between two new Moons and, as we saw in 'Why do you sometimes see the Moon during the day?', it's a little more than 29-and-a-half days. Over this period, every point on the Moon's surface (other than the polar regions) experiences a sunrise and a sunset, just as we do on Earth.

A related and rather more interesting question is *why* does the Moon always keep the same face towards the Earth? The answer to that goes back to something discussed in Chapter 3 – the effect of tides. We saw that the Moon's gravitational tug is what raises tides in the Earth's oceans, moving large bodies of water and pulling them through narrow coastal channels and straits. This dissipates energy in a process called tidal friction which, in the case of the Moon acting tidally upon the Earth, tends to slow down the Earth's rotation. The Earth also raises tides on the Moon – although in solid rock rather than water – and so, over the 4.6 billion-year history of Earth–Moon system, a similar tidal friction has acted to slow down the Moon's rotation.

Because the Moon's mass is only 1 per cent of that of the Earth, this latter process has already reached its ultimate conclusion. The Moon is now tidally locked into a state known as 'synchronous rotation', in which it always keeps the same hemisphere towards the Earth. Several other bodies in the Solar System experience synchronous or near-synchronous rotation, including Mercury, Venus and most planetary satellites.

What is meant by the dark side of the Moon?

For many people, it means the immortal Pink Floyd album of 1973, which reached number two in the UK album chart and remained listed in the Top 50 for most of the remainder of the decade. No wonder so many of us make that association. In astronomical terms, however, it simply means the side of the Moon that is not illuminated by sunlight. As we saw in the previous answer, most places on the Moon's surface experience alternate day- and night-time, just as we do on Earth – only rather more slowly.

There is another common, but incorrect, interpretation of the expression. Because the Moon always keeps the same face towards Earth, it has one hemisphere that is permanently hidden from our view, which many people refer to as the 'dark side of the Moon' – even though for half the month it is illuminated by the Sun.

In fact, the Moon's hidden portion is slightly *less* than a hemisphere, because the Moon nods slightly from side to side – and up and down – as it progresses around its orbit throughout the month, a phenomenon known as 'libration'. The effect allows us to see 59 per cent of the lunar surface from Earth, and is dramatically illustrated in a time-lapse movie sequence that can be found at http://antwrp.gsfc.nasa.gov/apod/ap010218.html.

Is it true that the Moon is receding from the Earth?

Yes, it is. At the moment, the Moon is drifting away from us by 3.82 centimetres per year. How do we know? By a technique known as lunar laser ranging, in which a laser beam is aimed at reflectors left on the Moon's surface by the *Apollo* astronauts. The returning beam is timed to give an exact distance. Many such measurements obtained over several years give us the accurate figure quoted above.

Once again, the effect is due to the tides raised by the Moon on the Earth, but in this case the exact mechanism is rather a subtle one. As we saw in Chapter 3 ('Why are there two high tides per day?') the effect of the Moon's gravitational pull is to raise a tidal bulge in the oceans of the Earth (and, to a lesser extent, in its solid crust). You might expect that bulge to be aligned with the direction of the Moon – and indeed it would be, if the Earth wasn't rotating. But as our planet turns on its axis, tidal friction (see 'What's the length of a day on the Moon?') drags the bulge along with it. This produces a slight offset of the bulge in relation to the direction of the Moon.

Curiously, the misaligned bulge now exerts a small gravitational pull of its own on the Moon, tending to speed it up very slightly in its orbit. If you've read Chapter 5, you'll know that speeding up a satellite causes it to move away from the Earth, and this is why the Moon is doing exactly the same thing. Of course, the effect is so small compared with the lunar orbit that it will take a hundred million years for the Moon's distance to increase even by 1 per cent.

The longer-term outcome of this process is rather more interesting, though. As we have seen, not only is the Moon drifting away from us, but the same tidal friction that causes the drift is also slowing the Earth's rotation. So, on a timescale of billions of years, the system will stabilise. The Moon will take approximately 47 days to orbit the Earth at a distance of rather more than half a million kilometres, but the planet's rotation will also have slowed to 47 days. Thus, our day will last as long as a month, and the Earth and Moon will perpetually face one another. A rather different situation from what we are used to – and one that our distant descendants will no doubt find cause to complain about...

MOONSTRUCK: EARTH'S ESSENTIAL SATELLITE

What would be the effect on Earth of the Moon not being there?

This is a very intelligent question, and one with several possible answers. I think what the listener had in mind was what the effect might be of the Moon suddenly being removed. And the answer to that is that if we could cope with the gravitational disturbance of our satellite suddenly being whisked away (which is doubtful), it wouldn't make much difference. We would no longer have the tides we have now (although the Sun would still have some influence), and the nights would always be dark, but that would be about it. I, for one, would mourn the absence of eclipses – there's nothing quite like an eclipse for concentrating people's attention on the Universe. And, well, the night sky would also be a lot less romantic without the Moon coming and going.

A much more interesting aspect of this question is what would be the effect of the Moon *never* having been there? Most astrobiologists (scientists who study the evolution of life in a cosmic context) agree that it's highly probable that human life would never have evolved on a moonless Earth. One reason for this is the stabilising effect our Moon has on the tilt of the Earth's axis. Without the Moon's substantial mass acting as a kind of flywheel, the Earth would have flipped many times during its history, causing chaotic disturbances to its evolving climate and inhibiting the development of complex biological organisms.

Moreover, the tidal effect of the Moon on the Earth's oceans may well have played an important role in the development of land animals. That twice-daily flooding of rocky shorelines could have eased the transition from sea to land for primitive organisms in a way that a non-tidal environment would have failed to do. So perhaps the pale orb of our satellite is the main reason why we have evolved on dry land rather than in Earth's more extensive oceans. And why we call our planet 'Earth' rather than 'Water'.

Why does the Moon appear to be the same size as the Sun in the sky?

Of all the questions in this book, I think this is one of the most provocative. The usual answer given in astronomy texts is that it is nothing more than a remarkable coincidence. But there are some rather subtle facts hidden here. First, the discs of the Sun and the Moon are among the most perfectly circular of all the objects in the Solar System – unlike other rotating bodies such as the Earth, which are slightly flattened at their poles. Second, while the Sun is some 400 times bigger than the Moon, it's also about 400 times further away, so both objects appear to be half a degree in diameter on the sky. In fact, because the orbits of both the Earth and the Moon are somewhat non-circular, the Sun and Moon appear to vary slightly in size, but their average 'angular' diameters are almost exactly the same.

The most significant consequence of all this is the occurrence of total solar eclipses. Everyone knows what happens – the new Moon passes exactly between the Sun and the Earth, and the dark centre of the Moon's shadow traces a narrow path across our planet's surface ('the path of totality'). The maximum width of the path is 270 kilometres, and the Moon's shadow races along it from West to East at a brisk 3,200 km/h. Everywhere along the path, the Moon's disc is seen to obliterate the Sun, allowing the faint outer atmosphere of our star to become visible, and attracting crowds of spectators. Depending on the circumstances of the eclipse, its duration at a particular location can be anything up to 7 minutes 31 seconds, a maximum determined by orbital mechanics and geography. In some eclipses, the Moon's position in its non-circular orbit is such that it appears slightly smaller than the Sun, and the obscuration is not complete. A ring of sunlight then remains around the Moon's disc, and such an eclipse is described, appropriately enough, as 'annular' (ringlike).

It is an extraordinary combination of circumstances that gives us this awe-inspiring, visually stunning and scientifically useful phenomenon. But let's just play devil's advocate for a moment. Suppose you are the Creator, and you want to make a sign to humankind that you put this Universe together. Wouldn't you choose something that anyone, anywhere on Earth could see without any kind of aid – something that would turn up periodically, just to remind your people that you were still around? And wouldn't you make it something so incredibly spectacular that absolutely no one could miss it? What better way than to turn the Sun into a terrifying black disc from time to time? That would certainly scare the pants off most primitive peoples – and clearly did, perhaps contributing to the rich diversity of religions. Frankly, I'm amazed that today's Intelligent Design Lobby hasn't jumped on the apparent diameters of the Sun and Moon to support their case that some things don't just happen by chance.

When scientists look at a phenomenon like this, they apply a philosophical device known as Occam's Razor, which basically says that you should adopt the simplest explanation. This would never involve an appeal to a supernatural being, since such a being is not scientifically verifiable and clearly strays from Occam's principle. (He was William of Ockham, by the way, c.1288–c.1348). Occam's Razor therefore says that the similar apparent diameters of the Sun and Moon is just a coincidence, and needs no further explanation. But, in fact, there might be one. And rather a good one, too.

A few years ago, an American scientist called Guillermo Gonzalez noted some additional facts about total eclipses. First, the phenomenon of a satellite just being able to cover the Sun as seen from its parent planet is rare in the Solar System, occurring only in the case of the Moon, and Saturn's satellite Prometheus (which, however, being irregularly shaped and very fast-moving, produces extremely brief eclipses on Saturn). Second, total eclipses of the Sun are a temporary phenomenon. As we saw a few questions ago, the Moon is receding from the Earth, and so eclipses of the kind we see now will only occur during a window of about 300 million years – some 6 per cent of the age of the Earth. At present, we are roughly halfway through that window.

Is it just coincidence that humans should have evolved on the planet exactly at the same time as this phenomenon is on display? Gonzalez mounts an argument that suggests that it could be an inevitability. We know that the Earth occupies a special place in the Solar System, the so-called 'Goldilocks zone', where the temperature is not too hot and not too cold, but *just right* for liquid water to exist – that being perhaps the single most important requirement for life to evolve. Moreover, the Sun is a particular type of star with a long, stable life, and all these stars are about the same size. These two facts together suggest that seen from a planet inhabited by intelligent life-forms, the parent star will *always* be about half a degree in diameter.

As we saw in the previous question, we also know that the Earth's unusually large satellite has played an important role in stabilising its rotation. Therefore, to get humans, perhaps you need a moon about the size of ours whose orbit can evolve in such a way that at some point, it, too, will have a half-degree diameter, and therefore exactly cover the parent star in the sky. If the timescale for that to happen matches that for intelligent life to evolve – as it does in our case – then perhaps the two phenomena go together. Wherever you have intelligent life, you might expect to see total eclipses. It's a remarkable conclusion, and one that's still controversial. It also suggests – rather depressingly – that because of the particular set of circumstances required, intelligent life might be very rare throughout the Universe.

LUNATIC FRINGE...

Did humans really walk on the Moon?

Well, according to a 2001 Fox television production called 'Conspiracy Theory: Did We Land On The Moon?', no, they didn't. This so-called documentary cited a number of supposed inconsistencies in the still and video imagery from the *Apollo* lunar landing programme, and suggested that the whole thing was staged in a movie studio to make it look as if the United States had won the space race hands-down. The sad part is that audiences were taken in, and I still get questions from people who wonder if we really did land on the Moon.

To be honest, asking a scientist if humans walked on the Moon is akin to asking a soldier if the Second World War or the Vietnam War happened. Researchers of all kinds – not just astronomers and space scientists – are immersed in the wealth of data that the *Apollo* missions provided. Twelve men walked on the Moon between 1969 and 1972, bringing back extraordinary first-hand accounts of their experiences. Mountains of detailed technical papers have been produced since. We have 380 kilograms of lunar rock and soil samples that are quite different from anything found on Earth. We have centimetre-accuracy distance measurements to the Moon, made using laser reflectors left on the surface by the astronauts. And we have 30,000 images of the lunar landscape, breathtaking in their timelessness and desolation.

It's quite easy to refute the TV programme's claimed inconsistencies. The explanations hinge on such aspects as the contrast between the bright lunar surface and the blackness of the sky (which is why no stars are visible in the photographs), and induced vibrations in a flagpole erected by the astronauts (which is why the flag appears to be flapping in one shot, despite the absence of an atmosphere). And so on.

But the best argument against the idea of a conspiracy is the most obvious one. Scientists and engineers are hopeless at keeping secrets, and tens of thousands of them worked on the project. It really would have been easier to send astronauts to the Moon than to keep all those talented and talkative people from spilling the beans.

CHAPTER 7

MORE THAN JUST EIGHT PLANETS

THE NEW SOLAR SYSTEM

I guess it might be considered rather fortuitous that these words are being written in the wake of the biggest upheaval in our view of the Solar System for three-quarters of a century. It's far better to be able to write with hindsight, of course. But it's not merely the demotion of Pluto to the status of dwarf planet (as determined by the International Astronomical Union at its General Assembly in Prague in August 2006) that leads me to speak of 'the new Solar System'. The Pluto episode, of course, was the public face of the revolution, and was greeted with something less than sympathy by the world at large. 'Pluto dumped by the über-nerds of Prague' was the headline I liked best. Not that I was at the meeting, but a lot of my über-nerd colleagues were.

Sadly, the main reason for public anguish at Pluto's reclassification seems to have been the fact that various handy mnemonics for remembering the order of the planets will no longer work. A number of those caught my eye, too, especially the one for indignant motorists in Sydney's fashionable north-shore suburbs:'My Very Expensive Mercedes Just Stopped Up Near Pymble'. To which, however, we can now counter 'My Virtuously Economical Mazda Jauntily Speeds Us Northwards'. *Touché*.

If the restoration of the Sun's family of planets to its pre-1930 status had been all there was to the issue, the press would have been right to be scathing. But in fact, this change is merely the tip of an iceberg of new knowledge about our Solar System, particularly in the furthest reaches of the Sun's domain. This revolution in understanding had been gathering momentum since 1950, when two American scientists – Kenneth Edgeworth and Gerard Kuiper – independently postulated the existence of an unseen asteroid belt in the Solar System's freezing outer reaches, beyond the orbit of Neptune.

Edgeworth and Kuiper imagined that this flattened disc of debris would constitute the leftovers from the formation of the Solar System 4.6 billion years ago, and predicted the existence of tens of thousands of small objects that had never coalesced together to build fully fledged planets. These objects would be very different in composition from the well-known asteroids in the so-called 'main belt', between the orbits of Mars and Jupiter. Never having been processed by heat, they would be icy agglomerates, resembling comets rather than the rocky bodies of the inner Solar System.

It was in 1992 that the first member of this new class was finally spotted, a remote object with the uninspiring name of 1992 QB1. More discoveries followed, and today more than a thousand have been catalogued, mostly with sizes ranging from 100 to 500 kilometres. They are usually known equivalently as Kuiper Belt objects (KBOs) or trans-Neptunian objects (TNOs) – although, technically, the Kuiper Belt is only one component of the trans-Neptunian region. But, with an apology to the purists, we'll quietly ignore that. The bottom line is that they're all a *very* long way from the Sun.

Rather closer to home, a growing understanding of the role of asteroid impacts in the history of the Earth – together with a few close encounters with so-called Near-Earth Objects (NEOs) – began to encourage eager scientists and force reluctant politicians to consider the risks posed by these objects, both today and in the future. The findings make sobering reading. Not because the risk is particularly high, but because the consequences of an impact by anything bigger than about 5 kilometres are global in scale, and could even eliminate humankind. Recognition of these risks led to a concerted effort by the USA to identify at least 90 per cent of all objects above 1 kilometre in diameter, but with funding that can only be described as meagre. When this study is complete, in the next year or so, it's likely that Congress will mandate NASA to hunt for smaller objects – perhaps down to 140 metres across. Hopefully, more substantial funding will be made available. These initiatives, together with research into how we might deal with an impact threat should one be identified, are expected to be included in a United Nations treaty on NEOs due to be presented in 2009.

The first asteroid was discovered back in 1801, and the new object was immediately hailed as the eighth planet of the Solar System (Neptune still being unknown). It was christened 'Ceres' after the Roman goddess of fertility. As time went by, however, it quickly became apparent that there was very little that was fertile about Ceres. It was a tiny world. Very soon, other small objects turned up in the same neck of the Solar System – leading quickly to the idea that these were not planets after all, but something new. It was left to William Herschel, the elder statesman of British astronomy, to coin the term 'asteroid' and give them all a decent identity.

The recent process in which Pluto lost its status as a planet bears a striking similarity to this episode – only on a much longer timescale. When Pluto was discovered in 1930 after a lengthy search, it was thought to be the missing link that would explain observed irregularities in the orbit of Uranus. As time went by, however, it gradually became clear that Pluto was too small to have any effect on the giant seventh planet and, as successive estimates of its diameter became ever smaller, its status shrank accordingly. We now know that Pluto is only two-thirds the size of the Moon and, moreover, has an orbit quite different from those of the other planets – very elongated, with a 17-degree tilt to everything else. To add insult to injury, the need to explain irregularities in Uranus' orbit evaporated during the 1980s, when data from *Voyager 2* allowed scientists to revise Neptune's mass.

So what was to be done about Pluto? It had been a planet for so long that to demote it to something inferior seemed unthinkable. On the other hand, it had been obvious since the early 1990s that Pluto wasn't a planet in the usual sense of the word – even though it boasted a decent-sized satellite, Charon, discovered in 1978. (It's pronounced 'Care-on', by the way, not – *please* – 'Sharyn'.) Pluto is also now known to have two smaller moons, Nix and Hydra. Inevitably, the spotlight was turned on to exactly how we define a planet. And here, the Pluto question conspired with a plethora of other new discoveries to cause real anguish.

Only a couple of decades earlier, there would have been no need to ask what constituted a planet. The dictionary definition was perfectly adequate: 'A large heavenly body orbiting the Sun or another star and shining only by reflected light'. The first hint of a problem arose when the *Pioneer* and *Voyager* spacecraft sent back accurate measurements of Jupiter. It turns out that the giant planet emits 1.7 times more heat than it receives from the Sun, and therefore – at least in the infrared region of the spectrum – 'shines' by its own light rather than reflected light. Perhaps an arcane point, but a portent of things to come.

As we have seen, the first KBO was discovered in 1992, blurring the distinction between Pluto and this large class of icy asteroids. Then, in 1995, the first planet orbiting a normal star other than the Sun was unequivocally detected. Like many that have been discovered since, it is an object comparable in mass with Jupiter but in an orbit so close to its parent body (a Sun-like star named 51 Pegasi) that it almost skims the surface. It is very unlike our normal idea of a planet.

The last few years have also seen the discovery of several free-floating planets – objects whose mass is less than 13 times that of Jupiter (above which they would be classed as weakly shining stars called brown dwarfs), but which are orphaned in deep space with no parent star. How these objects came into being remains a mystery. Finally, in 2003, the inevitable discovery of a KBO bigger than Pluto brought a new level of urgency to the need for a comprehensive definition of a planet. This catalyst object was originally called 2003 UB313 (and subsequently nicknamed Xena), but is now formally named Eris, after the Greek goddess of strife and discord. Very fitting, given the controversy that surrounded its classification. Eris is likely to be merely the first-discovered of many distant bodies larger than Pluto, as our ability to detect them grows with the size of our telescopes.

The International Astronomical Union's response to all this was predictable. They formed a committee. Its brief was to come up with a comprehensive definition of a planet, taking input from many individual astronomers. This they did, but it was eventually left to an open vote at the Prague meeting to determine the outcome. What were the options? Those that relate particularly to our Solar System included:

1. To define planets as Sun-orbiting spherical objects – that is, those large enough (more than about 700 kilometres across) for their own gravity to pull them into a spherical shape. This is a straightforward definition, eliminating most asteroids and KBOs, but admitting Ceres and the largest KBOs. A related idea was to set an arbitrary lower limit on diameter (2,000 kilometres was suggested), so that out of the Solar System's smaller bodies only Pluto and Eris would squeeze in.
2. Something like the above, but with the added proviso that a planet is only a planet if it is the dominant object in its own region of the Solar System. That would effectively eliminate *all* asteroids and KBOs, including Pluto and Eris.
3. An even more arbitrary idea was to invoke tradition and decree that Pluto should retain its planet status out of deference to history. Thus, the status quo would remain.

As we now know, option two won the vote. The first option would have given us a Solar System with something like a dozen planets, making the search for suitable mnemonics quite a challenge and, more seriously, raising the possibility of an ever-increasing suite of planets as new discoveries were made. The third option was felt to be a bit wimpish, one that would leave scientists open to the criticism that they do not confront the discovery of new knowledge in a rational way. But the winning option did leave a loose end. To tie it up, the International Astronomical Union took the further step of defining a new class of object – dwarf planet – which includes Pluto, Ceres and Eris. That's how Pluto was dumped as a planet. Putting sentiment aside, it was, in the view of most astronomers, the correct decision.

So, after all these monumental changes, what does the current inventory of the Solar System look like? The largest and most important object remains the Sun – our star, whose formation spawned everything else as by-products. Then there are the four gas giants – Jupiter, Saturn, Uranus and Neptune – whose dense, opaque atmospheres constitute most of their mass, and hide any compact solid core that may lurk at their centres. The four rocky planets are next – Earth, Venus, Mars and Mercury. Then the handful of newly defined dwarf planets, of which Eris is currently the largest known. With the exception of Mercury and Venus, all the planets are known to have satellites orbiting around them – as do many of the Solar System's smaller bodies, such as asteroids and KBOs.

The asteroids themselves number more than a million, mostly in the main belt, although other populations exist. The distant KBOs also divide into sub-populations, including some known as 'Centaurs', which have similar characteristics to comets. (Half-man, half-beast; half-KBO, half-comet. Who says astronomers have no soul?) Then there are the comets themselves, mountain-sized orbiting snowdrifts that become luminous as they approach the Sun and its radiation causes their ice to evaporate. Comets are thought to originate in a spherical halo surrounding our star at a distance roughly halfway between it and the next star. It is known as the Oort Cloud, after the eminent Dutch astronomer who first postulated its existence.

Finally, the space between the planets is inundated with dust, subatomic particles and twisted magnetic fields – not to mention a fair number of small metallic objects that have emanated from the largest of the rocky planets. While our spacecraft are clearly not naturally occurring bodies, many of them do now form a permanent part of the Sun's retinue. Altogether, then, the Solar System is a *very* complex place. One might almost describe it as a mess. It is a far cry from the neat and tidy planetary system we thought we lived in when I was a lad – but it's a heck of a lot more interesting.

A TOUCH OF SUNBURN: THE FURNACE AT THE CENTRE OF THE SOLAR SYSTEM

How does the Sun burn without oxygen?

It's almost the first science lesson you learn at school. Without oxygen, combustion cannot take place. Yet here's this fiery object in our skies, apparently burning away happily in the vacuum of outer space. How does it work? If you're worried that this is a dumb question, take heart. It was only during the last century that the energetic processes occurring within the Sun were finally understood.

Although it had long been suspected that the Sun was just a star seen close by, until the second half of the nineteenth century no one had any idea how it worked. Then, a number of physicists suggested that stars' energy (and that of the Sun) came from the heat generated by the collapse of a ball of gas under its own gravity. It's a neat idea, analogous to the well-known experiment in which a tyre pump gets hot as you inflate the tyre. Compressed gas heats up, whether the compression is due to tyre pumping on Earth or gravitational collapse in space. And indeed, today we know that this heating process is crucial in the early stages of the formation of a star.

However, in 1920, the great English astrophysicist Arthur Eddington demonstrated that this explanation was quite inadequate to account for the typical lifetime of stars – including the Sun. They would simply cool down and, well, go out. Eddington suggested that perhaps nuclear reactions, then being explored by Ernest Rutherford in Cambridge, were responsible for the energy output. 'What is possible in the

Cavendish Laboratory may not be too difficult in the Sun,' he remarked. He always did have a way with words.

It was left to other physicists to calculate the exact nature of those nuclear reactions, most notably Hans Bethe in 1939. Starting with hydrogen, the most basic and abundant raw material in the Universe, atoms (or rather their cores, or nuclei) would fuse together to make other elements, simultaneously releasing energy in the form of photons. Essentially, these are particles of light – but in this case, they are high-energy gamma ray photons. As they zip through the Sun's interior, they are eventually converted into the heat and light radiated from the surface. In the Sun, the process of nuclear fusion takes place deep in our star's heart at a temperature of around 15 million degrees, and its main by-product is helium. The Sun is literally a nuclear furnace.

These reactions are self-sustaining and, once started, will continue until the fuel runs out. There is no need for oxygen; it, in fact, is another by-product of the reactions themselves.

When will the Sun stop burning?

Every second, the Sun uses up four million tonnes of hydrogen, converting it primarily into helium and energy. By our standards that is a prodigious rate of fuel consumption, one which seems quite alarming at first sight. But the Sun is a *very* large object, and its mass is so great that there is enough hydrogen to last for some ten billion years. At present, the Sun is roughly halfway through its lifetime, so the hydrogen won't run out for about five billion years. What will happen then?

Such is our knowledge of the lifecycles of stars, gained over a century and a half of astrophysics, that we can predict what will happen with a high degree of certainty. Unfortunately, it's not pleasant. The Sun will swell rather spectacularly. We know that the size of a star is controlled by two opposing forces – the so-called radiation pressure produced by nuclear fusion in the central core (see previous answer), which pushes outwards, and the pull of gravity, which tries to make the star collapse. When that balance is altered, the diameter of the star will change.

In the Sun's case, as its central core runs out of hydrogen it will shrink, becoming hotter and denser. Eventually, the nuclear reactions change, and the core will begin burning helium to make other elements – principally carbon. Meanwhile, the Sun's outer layers have compensated for the core's collapse by growing to enormous proportions, becoming extremely rarefied in the process. These outer layers are relatively cool, around 2,000– 3,000°C (compared with the present surface temperature of 5,500°C), and the Sun has become a 'red giant'. At this stage, it is quite likely that the Sun will have swallowed the Earth, such will be its swollen proportions.

Other processes follow on from this, with the Sun's outer layers blowing off in the form of a dense wind of particles, allowing radiation from the naked core to sweep up the wind into a shell of material. We see many glowing objects of this kind when we look deep into space; they are known as planetary nebulae because of their planet-like appearance, and can be extremely spectacular. ('Nebula', by the way, just means 'mist', and was used by early astronomers to describe any misty patch in the sky.) Eventually, the nebula blows away, leaving only the hot, dense core. The Sun has now become a white dwarf – of which more in the next chapter.

All this talk of the Sun's demise reminds me of the one joke that astronomers like to tell at parties. (Don't get excited – it's pretty ordinary.) To be technically accurate, it's not so much about astronomers as astrophysicists, who study the physics of the Universe and everything in it. So – two astrophysicists are sitting on a bus. One says to the other 'Did you know that the Sun will become a red giant star in five billion years' time?' 'Yes, I did,' replies the other. 'Excuse me,' says a man sitting behind them, 'what did you say?' 'Well,' says the first astrophysicist, 'I was just saying that the Sun will become a red giant star in five billion years.' 'Oh, thank God for that,' says the man. 'For a moment I thought you said five *million* years.' Yes, folks, it's just as well astronomers don't have to earn their living telling jokes.

What are sunspots, and do they have any influence on Earth's climate?

Although sunspots occasionally can be big enough to be seen with the naked eye (and, if you didn't know already, it's extremely dangerous to look at the Sun without special filter lenses), it was only Galileo's legendary exploitation of his new telescope, beginning in 1610, that made them common knowledge. Two centuries later, William Herschel thought they were windows in the Sun's fiery atmosphere that allowed its cool, rocky, and possibly inhabited surface to be seen. Not his best idea. Today, we know that they are areas of the solar photosphere (its visible 'surface') where the temperature is up to 2,000 degrees lower than that of the surrounding gas. Hence, they appear darker.

Modern observations of sunspots using ground- and space-based telescopes reveal that they are astonishingly complex regions of intense magnetic activity. They are associated with coronal mass ejections and solar flares (see 'How does the Sun affect communications and power transmissions on Earth?' in Chapter 3.) Sunspot numbers are greatest at the peak of the solar cycle – which, despite its name, is not an environmentally friendly motorised bike, but a regular, 11-year variation in the Sun's level of activity.

A link between the level of solar activity and the Earth's climate has long been suspected. The earliest evidence came from something called the Maunder Minimum, named after a chap called Walter Maunder who described the effect in 1922.

Maunder had established from various records that sunspots appeared in only minimal numbers between the years 1645 and 1715. But that period also sat right in the middle of an era of unusually low temperatures in European countries, often referred to as the Little Ice Age. Very suspicious.

What has eluded scientists trying to link solar activity with the Earth's climate has been a mechanism to produce the effect. We saw in Chapter 3, however, that the Earth is constantly bathed in the solar wind, a magnetised stream of subatomic particles emanating from our star. The solar wind's intensity is closely connected with the level of solar activity, and a new generation of spacecraft (led in 1995 by SOHO, ESA's Solar and Heliospheric Observatory) has analysed its behaviour in detail. At the terrestrial end, we now know that there are links between cloud formation and the bombardment of the atmosphere by subatomic particles. Thus, while the details are far from complete, it does seem likely that solar activity – as manifested by sunspots – can, indeed, influence the Earth's climate.

PLANETS' PROGRESS: WHY THE SOLAR SYSTEM LOOKS LIKE IT DOES

How did the Solar System originate?

'Ah, der hard ones firsht' – as Eccles used to say in 'The Goon Show' (on being asked his name...). It's certainly true that, in big picture terms, the formation of the Solar System is well understood, but there are still areas of detail that remain uncertain. Some are quite controversial, and competing hypotheses for the formation of planets and satellites (and how they came to be where we find them today) can be debated fiercely.

Did you know, by the way, that the study of the origin and evolution of planetary systems is called cosmogony? It's not a common term, but the science of cosmogony has received new impetus over the past few decades as spacecraft exploring the Solar System have sent important new data back to Earth. These data – together with the ever-increasing capacity of modern computers to perform large-scale feats of number crunching – have allowed cosmogonists to develop increasingly sophisticated models of planet formation. Even more significantly, the discovery during the past ten years or so of more than 250 'extrasolar' planets (which orbit stars other than the Sun) has provided a wide range of scenarios that must be allowable in any model of planet formation. Most of those planets are in 'solar systems' very different from ours, but presumably have been formed by similar processes.

Planets are formed as by-products of star formation, and it's no exaggeration to suggest that they are merely the cinders left over from that process, since the central star is always the most important entity – at least in terms of mass. Good thing for us that left-over cinders seem to be an inevitable consequence.

Star formation starts with a large cloud of gas (mostly hydrogen and helium), polluted by grains of dust made of heavier elements such as oxygen, silicon and iron. While the dust is a product of earlier generations of stars, the hydrogen and helium are primordial, having originated in the Big Bang. Usually, the cloud of gas and dust is large enough that many stars will form in it, eventually creating a star cluster of the kind in which our Sun undoubtedly was born. Each star begins with a local collapse of the cloud under its own gravity. This can be triggered by a variety of events such as changes in the underlying magnetic field, or gravitational disturbances due to another gas cloud passing by, but the end result is the same – as the gas is compressed its temperature rises until, after a few million years, we see the onset of nuclear reactions as described a couple of questions ago.

The dust becomes a 'protoplanetary disc' of material swirling around the baby star (or 'protostar'), adopting this flattened shape due to its rotation. It is within this disc that planet formation takes place, and the basic process is one of accretion, in which the dust grains stick together to make successively larger particles. In the turbulent central plane of the disc, collisions between particles are frequent, leading eventually to the growth of kilometre-sized bodies known as planetesimals. At this stage, collisions result in both the growth of objects and their fragmentation. Growth is the dominant effect, however, and over millions of years successively larger 'protoplanets' emerge, fewer in number and more widely spaced than their planetesimal forebears.

By now, the protoplanetary disc is starting to look much more like a 'solar system', but the process of sweeping up the debris into planet-sized objects goes on for tens to hundreds of millions of years. Indeed, in our Solar System, it is still going on at a residual level, as witnessed by the thousands of tonnes of meteoroidal material that hit the Earth every year.

Although accretion can be considered the bare bones of the planet formation process, even a cursory glance at our Solar System makes it clear that there is much more to it than this. For example, the four rocky planets (Mercury, Venus, Earth and Mars) all lie within the main asteroid belt, and are quite different from the four gas giants (Jupiter, Saturn, Uranus and Neptune) that lie beyond. The suggestion is that this is because the dust grains nearer to the youthful star are heated to high temperatures, and may even be vaporised in the early stages of the process, leading to a modification of the accretion mechanism.

Further from the protostar, cooler temperatures allow emerging protoplanets to grow by collecting gas from the remnants of the collapsing cloud, ultimately becoming gas giants. Any residual gas, by the way, is eventually blown away by strong winds of subatomic particles from the new-born star as it goes through something called the T Tauri stage. (I guess it's a bit like colic in a human infant.) The would-be giant planets may also accrete icy planetesimals of the kind found today in the cold outer reaches of the Solar System – the Kuiper Belt objects. That process is thought to have been particularly important in the growth of Uranus and Neptune.

The extent to which our Solar System is representative of normal planet formation processes remains an open question, because we are not yet able to study extrasolar planets in sufficient detail to make valid comparisons. It may be, therefore, that some aspects of our planetary system are quite untypical. For example, recent studies suggest that dust grains found in meteoritic material are rounded, as if they have been seared by high-energy radiation from an event such as a nearby supernova (exploding star). If these rounded dust grains are better able to stick together than unheated ones, it may be that a nearby supernova at a particular stage in the process is an essential ingredient for the formation of rocky planets. Such ideas remain unproven, however.

One further aspect of the Solar System seems to need at least some attempt at an explanation in any theory of cosmogony. This is a curious numerical progression known as Bode's Law, formulated in 1772 by the German astronomer Johannes Bode, although it had been noted some years earlier by one Johann Titius. If you take a doubling sequence of numbers starting with zero (0, 3, 6, 12, 24, 48, 96, 192,...) and then add a four to all the numbers, you get 4, 7, 10, 16, 28, 52, 100, 196,... Now, guess what? Taking the Earth's distance from the Sun as 10, this is a fairly good representation of the average distances of the other planets from the Sun: Mercury 3.9, Venus 7.2, Earth 10.0 (of course), Mars 15.2, Ceres (the largest asteroid) 27.7, Jupiter 52.0, Saturn 95.4 and Uranus 191.8. The next step should be

388 and, although Neptune (300.7) doesn't fi t the progression, Pluto (394.6) just about does.

Early theorists of the origin of the Solar System went to great lengths to try to explain this effect, although it seems to have become unfashionable these days, with most scientists regarding it merely as a numerical curiosity that has no underlying physical significance. My bet, however, is that one day, some subtle aspect of planet formation theory will reveal the true significance of Bode's Law, and the grand old German will be vindicated. Watch this space.

Why do all the planets rotate and not slow down?

Back in the 1800s, when I was at school, I remember being told the difference between an object that was rotating and one that was revolving. Planets, for example, rotate on their axes, but revolve around the Sun in their orbits. So – if something that spins on its axis is rotating, why isn't a revolver called a rotator? 'The Arizona Kid pulled out his rotator, and drilled the sheriff full of holes.' Somehow, it conjures up an image of a cowboy armed with a motorised cultivator.

The relevance of this piece of nonsense is that, seen from above its North Pole, the Earth both rotates and revolves in an anti-clockwise direction. Astronomers have an adjective for such motion – 'direct' (or, more rarely, 'prograde'). Motion in a clockwise direction, on the other hand, is termed 'retrograde'.

As it turns out, almost everything in the Solar System revolves and rotates in direct motion. Even the Sun spins on its axis in direct fashion, taking 25.4 days to complete one rotation (but slightly more at its poles). The few exceptions to the 'everything-goes-direct' rule are Venus, Uranus and Pluto, which have retrograde rotational motion (although Venus is almost synchronous – see 'What's the length of a day on the Moon?' in Chapter 6). Also, several of the outermost satellites of the gas giants orbit their planets in a retrograde direction, indicating that they are most likely captured objects, rather than satellites that formed with their planets.

The fact that almost everything in the Solar System goes around in the same direction seems hardly likely to be a coincidence and, indeed, it isn't. What we are seeing here is the fossilised rotation of the cloud of gas and dust from which the Solar System originally formed (see the previous answer). That cloud would have been slowly turning from the outset but, as it collapsed under its own gravity, this rotation would have been speeded up by a process called the conservation of angular momentum. It's an effect you can easily demonstrate by spinning yourself around on an office chair, first with your legs outstretched, and then with them pulled in towards the chair's axis. The increase in speed you feel is exactly what happened to that collapsing gas cloud some 4.6 billion years ago. It's a lot of fun, but if you do it too much you'll throw up – just as the protoplanetary disc threw up a bunch of rotating planets...

Why doesn't the rotation slow down? Well, actually, it is slowing down, but very slowly. In Chapter 6, we saw that the Moon's rotation has already slowed to match its period of revolution around the Earth. Due to tidal friction, the same process is taking place throughout the Solar System. However, there is so much rotational energy stored in the Solar System that the spin-down period is measured in tens of billions of years, so there is no need to panic yet.

Why do the planets lie more or less in a plane?

Going back two questions to the origins of the Solar System, you will recall that the early Sun spawned a rotating protoplanetary disc, in which the planets formed. The overwhelming tendency of material within that disc would be to follow gravity's pull towards its central plane, so that as the disc evolved it would become thinner and more well defined. It's the last vestige of this disc that we see in the planets' orbits today. Except for Mercury and Venus, the orbits of the planets all lie within two-and-a-half degrees of tilt to the Earth's. The fact that the two innermost planets have more of a tilt (Mercury with 7 degrees, and Venus with 3.4 degrees) is also consistent with the idea that they formed from a thin disc of material.

It was the tilt of Pluto's orbit to the Earth's (17 degrees), together with its elongated form, that first marked it out as being different from the other planets. That moderate tilt is fairly typical of KBOs, of which Pluto is now known to be a large example. In fact, during the 1980s, before the first KBO (other than Pluto) was discovered, the possibility of their existence was reinforced when the Kuiper Belt was identified as the probable source of short-period comets. These comets have orbital periods of less than 200 years and tend to have orbits reasonably close to the plane of the Solar System.

Not surprisingly, perhaps, given its great distance, the Kuiper Belt continues to throw surprises at us. Towards the end of 2005, an announcement was made of the discovery of a largish (500 to 1,000 kilometres across) KBO whose official name sounds like the latest in high-performance sports sedans – 2004 XR 190. Its Canadian discoverers, however, nicknamed it 'Buffy', after the eponymous TV vampire-slayer. Why? Because they thought it was going to be 'a bit of a theory-slayer'.

Buffy orbits in an unusually circular path at the further edge of the main concentration of KBOs, some 8.5 billion kilometres from the Sun, but what is really strange is the tilt of its orbit to the plane of the Solar System – a massive 47 degrees. Explaining how it got into an orbit with such a unique combination of circularity and tilt has proved difficult, and involves speculative ideas such as changes in the orbit of Neptune and encounters with nearby stars. It may even cause a rethink of the shape of the Kuiper Belt itself. Once again – watch this space.

What makes planets round?

At the beginning of this chapter, we noted that roundness (or, more accurately, sphericity) was suggested as one of the prerequisites for planetary status. Not that most planets are perfectly spherical, by the way; their rotation is usually great enough to cause a slight bulging at the equator. In the Solar System, the most notable example of this so-called 'oblateness' is found in the planet Saturn, whose polar diameter is some 10 per cent less than its equatorial diameter.

Notwithstanding oblateness, the mechanism by which planets become spherical is simply good old gravity. Above a certain size (usually taken to be about 700 kilometres across), a planetary object will have enough self-gravity to overcome the natural resistance of its rock to deformation. This means that the body will assume the only shape that is stable with respect to its own gravity – a sphere. You may recall images of the interiors of orbiting spacecraft that show droplets of liquid taking on a spherical shape. That happens for the same reason, only in this case the force is surface tension.

Gravity plays a further role in shaping planets, through a process known as 'differentiation'. In large planetary bodies such as the early Earth, the interior is subject to heating by radioactive processes, gravitational compression and the effects of impacting planetesimals. These combine to cause melting, allowing the denser materials to sink towards the centre of the planet and creating a core-mantle structure.

There is some evidence that large KBOs such as Pluto and Eris might be the products of incomplete planet formation. Here, the process seems to have been interrupted in mid-flow, resulting in half-finished worlds that have nevertheless become big enough for gravity to pull the solid material to the middle. The process of differentiation is likely to have given Pluto – and perhaps Eris too – a rocky core with an icy mantle. Pluto's surface is known to consist of frozen nitrogen, along with methane, carbon dioxide and ethane. However, the bulk of Pluto's icy mantle is likely to consist of water ice, buried beneath the more volatile surface ices. The planet's tenuous atmosphere, whose existence was confirmed in 1988, is probably mostly gaseous nitrogen.

Computer simulations of planet formation, carried out at institutions such as the Southwest Research Institute in Boulder, Colorado, demonstrate that Earth-sized objects could, in fact, have formed in the outer regions of the Solar System. Why the process stopped is a mystery. But if Pluto really is a half-built world, the close-ups that the *New Horizons* spacecraft will be sending us in 2015 will give us a unique opportunity to see planet formation in freeze-frame, providing new detail in our current understanding of the process.

What are Saturn's rings made of?

When Galileo embarked on his first spree of discovery with his newly perfected telescope in 1610, one of the heavenly bodies that attracted his attention was the planet Saturn. He noted privately that its appearance through the telescope was unusual, resembling a disc with two smaller ones touching it on either side. Despite further study, he never managed to figure out what was going on, and other puzzled astronomers of the time described Saturn as having 'ears'. Almost half a century would pass before the Dutch mathematician Christiaan Huygens realised that Saturn was, in fact, a planet with a broad ring around it – or, more exactly, a system of rings.

Just two centuries later, in 1859, the great Scottish physicist James Clerk Maxwell used nothing more than mathematics to demonstrate that Saturn's rings could not be solid and still retain their stability. He suggested that they must be made of millions of small bodies, each in an individual orbit around the planet. And so it turned out to be when, in 1895, an American observer called James Keeler directed that magical tool of astronomy, the spectroscope, towards Saturn's rings. The tell-tale rainbow spectrum allowed him to work out the speed of rotation at any point, and he found that sure enough, the rings were moving not as a solid body but as a swarm of particles.

The epic fly-by missions of the *Pioneer* and *Voyager* spacecraft in the 1970s brought us spectacular first-hand views of Saturn's ring system. These have now been supplanted by dazzling images from *Cassini*, which went into orbit around Saturn in July 2004. The new images and other measurements confirm that the rings consist of a complex series of ringlets – some curiously braided, others appearing to have spoke-like features, and many of them shepherded by Saturn's moons.

The rings appear so bright because they are highly efficient reflectors of sunlight. They consist mostly of water ice in chunks ranging in size from about a centimetre across to 10 metres or so. Astonishingly, the rings are no more than 100 metres thick and, with an overall diameter of around 275,000 kilometres, when seen edge-on they have a blade-like appearance. From our vantage point on Earth, they effectively disappear when that occurs. It will happen in 2009, and again in 2024.

The source of Saturn's rings is unknown, but they are most likely to have originated either at the same time as the planet, or in the disintegration of an icy comet nucleus. All four of the gas giant planets (Jupiter, Saturn, Uranus and Neptune) have a ring system of one sort or another, but none can compare in size or intensity with that of Saturn. Truly, it is the pearl of the Solar System.

PAS DE PLANETS: THE SOLAR SYSTEM'S ETERNAL DANCE

Are there any effects when the planets line up?

The complex dance of the planets in their orbits has fascinated us for as long as we have recognised that the Sun is at the centre of the Solar System. Perhaps that explains the popularity in earlier times of orreries – mechanical representations of the Solar System, with the planets geared to revolve around the Sun in the correct ratios of their orbital periods. These exquisite little machines are named after Charles Boyle, Fourth Earl of Cork and Orrery, for whom one was made early in the eighteenth century.

It is inevitable that as this dance progresses, there will be times when the planets will form something approaching a straight line (ignoring the relative tilts of their orbits – see 'Why do the planets lie more or less in a plane?'). Naturally, this listener's question then arises. It is a bit sad that the correct answer (which is 'No') is a lot less exciting than the scenarios portrayed in various scare-mongering publications over the years. Dire consequences have frequently been predicted to arise from planetary alignments, but they simply don't happen.

In fact, it's hard to see quite what mechanism the doom-and-gloom pundits imagine would give rise to a major catastrophe. I think the last one I heard about threatened a channelling of electromagnetic fields along the line of planets, ripping the Earth asunder in an apocalyptic cataclysm. Great stuff. In reality, the only force that acts significantly over these distances is gravity, and gravitational perturbations between one planet and another are taking place all the time. Believe it or not, the sum effect of these perturbations on the Earth during an alignment of the planets is incomparably smaller than that caused by the monthly lining up of the Sun and Moon.

If you don't believe me, cast your mind back to May 2000, when there was a reasonably tight alignment of the five naked-eye planets (albeit too close to the Sun in our skies to be observed from Earth). Did you feel any ill effects? No – I didn't think so. A tighter alignment of those same five planets (Mercury, Venus, Mars, Jupiter and Saturn) will take place after sunset on 8 September 2040, when, together with a slender crescent Moon, they will be within a 10-degree circle. I will be 95, and fully intend to be wheeled out in my bathchair by anyone who'll oblige – hopefully, a 35-year-old girlfriend – to have the spectacle pointed out to my aged eyes. Fact is, she'll be able to point out anything to me, and I'll be convinced. But at least I'll be able to die happy...

What is a conjunction? An opposition? A transit? An occultation?

These technical terms are the nitty-gritty of the dance of the planets, denoting the basic steps and flourishes. Conjunction is when two celestial objects – usually a planet and a star, a planet and the Sun or Moon, or two planets – lie in approximately the same direction in the sky.

If a planet is described simply as being 'in conjunction', then the other body is always assumed to be the Sun. For Mercury and Venus, there are two possible cases. Inferior conjunction is when the planet lies between the Sun and the Earth, and superior conjunction is when it is behind the Sun. Planets beyond the orbit of the Earth (Mars *et al.* – which, incidentally, are known as superior planets) can only be in superior conjunction. The two inferior planets (Mercury and Venus) also have significant events called 'greatest elongations', when they are at their maximum distance from the Sun in the sky, and thus most visible from Earth. Even at these points, the planets are no more than 28 degrees (Mercury) or 47 degrees (Venus) from the Sun.

The opposite situation to conjunction is opposition, when a planet is exactly opposite the Sun in the sky. It is visible all night, rising as the Sun sets and setting as it rises, exactly like a full Moon. Opposition is a status denied to Mercury and Venus (something most politicians would envy), but all the superior planets undergo it from time to time.

A little thought will show that opposition is the best time to observe one of these planets, for not only is it visible throughout the night, but it is also at its closest to Earth and illuminated face-on by the Sun. However, the fact that all the planets are in slightly non-circular orbits means that some oppositions are more favourable than others. Mars, for example, had a particularly close opposition at a distance of 56 million kilometres on 29 August 2003. Even though oppositions of Mars occur at intervals of slightly more than two years, the red planet won't be that close to us again for many millennia.

Transits of Mercury or Venus occur when the planet is in inferior conjunction and also crossing the plane of the Earth's orbit. At most of their conjunctions, Mercury or Venus pass above or below the Sun, as seen from Earth. During a transit, they move directly across it. Mercury transits occur 13 or 14 times per century, and the last one was on 9 November 2006. The next is on 9 May 2016 and, unlike the 2006 transit, it will be visible from the UK.

Transits of Venus are much rarer than those of Mercury, and during the eighteenth and nineteenth centuries these celebrated events were used to plumb the depths of the Solar System by means of a technique suggested by Edmond Halley. It involved the simultaneous observation of the transit from widely separated locations on the Earth. Venus transits occur in pairs eight years apart, but separated by more than a century. At the time of writing, we are between the two events of a pair – the transits of 8 June 2004 and 6 June 2012.

Finally, to 'occult' means simply to hide, and that's what happens during an occultation. A distant celestial object is hidden by a nearer one. The most common occultations involve the Moon as the occulting body – simply because it's relatively big and covers up a lot of sky as it moves along. Planets and stars are both frequently occulted by the Moon. More rarely, stars are occulted by planets and, even more rarely, planets (or their satellites) occasionally occult one another.

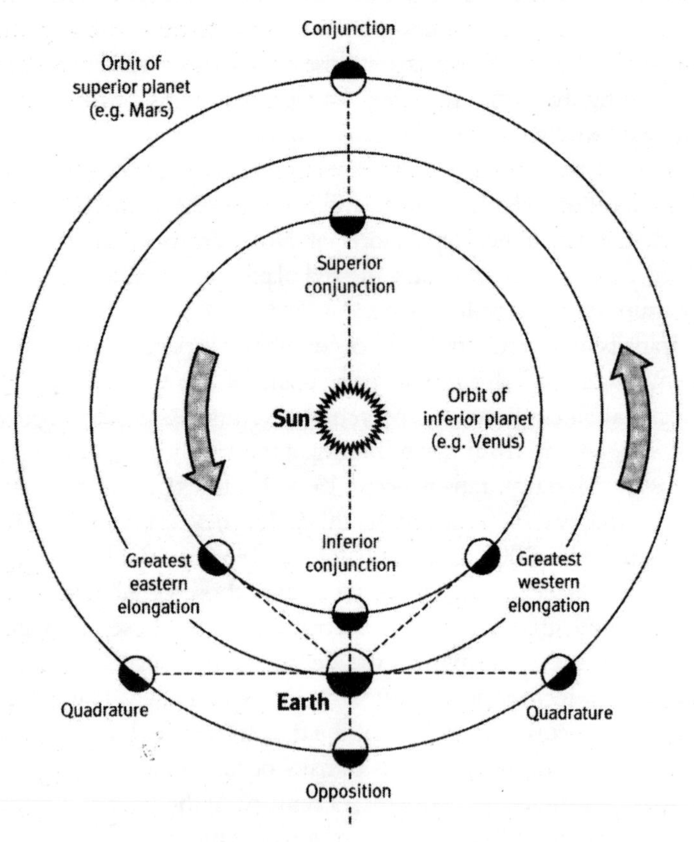

Planetary orbits as seen from above the Earth's North Pole

Red-letter days for inferior and superior planets.

Much can be learned from occultations. Accurate timings of lunar occultations by amateur astronomers have proved invaluable for calibrating both the Moon's orbit and its profile in terms of its mountain heights, and so on. Occultations by planets reveal such features as structure in their atmospheres (most recently in a 2006 occultation by newly demoted Pluto) and the presence of ring systems. The rings of Uranus, for example, were discovered during an occultation by the planet on 10 March 1977. Again, occultations by asteroids can show their edge-profiles and reveal whether they are accompanied by satellite asteroids.

The most spectacular occultations are those in which the Moon hides the Sun. They are more usually known as total solar eclipses, however...

What was the bright comet I saw in 1957, 1965, etc.?

Unlike the planets, which orbit the Sun in roughly circular paths, comets move in highly elongated trajectories that are sometimes not closed – that is, the comet will never return to the vicinity of the Sun.

We know that comets are weakly bound aggregations of ice and dust – flying snowdrifts – the size of a range of mountains. They probably represent material from the original gas and dust cloud that became the Sun, never having been processed by heat. It is this that has prompted the recent spate of space missions involving comets, such as *Deep Impact* (2005) and *Stardust* (2006). Of course, as soon as such 'cometary nuclei' get close to the Sun, they quickly develop the characteristic features of a comet – a coma (head), formed by the out-gassing of volatile materials and the release of dust, and a prominent tail. Sometimes, these features become very spectacular in our night skies, as in the 2007 apparition of Comet McNaught (C/2006 P1), discovered the previous August by my colleague

Rob McNaught at Siding Spring Observatory. A comet frequently appears brightest after its closest approach to the Sun (its 'perihelion passage'), especially if it is then close to the Earth.

The reason for today's great scientific interest in comets lies in what they might tell us about the formation of the Solar System, perhaps even about the origins of life on Earth. If the typical comet is a frozen remnant of the protoplanetary disc that surrounded the infant Sun, then its chemistry could be nothing less than the Rosetta Stone of our corner of the Universe, with pristine dust grains that have been forever cold, and organic ices that may contain the progenitors of living cells.

The importance of this to our planet's history is that impacting comets are thought to have been a significant source of ices such as water ice, methane and ammonia. It is highly likely that more complex organic molecules were included in the same packages, and a handful of scientists think that life itself may have arrived in this way. Hence the extraordinary interest in investigating comets' icy contents.

For most people, however, comets are occasional (and sometimes unexpected) visitors to our night skies that can punctuate a lifetime. That is why there are always listener questions to do with comets. I can entirely understand it. The appearance of Comet Arend-Roland (C/1956 R1) over the skies of northern England in late April 1957 was one of the little triggers that led a certain Fred Watson towards an all-consuming passion for astronomy. Interestingly, the twentieth century was far less notable for bright comets than the nineteenth, which included some bright enough to be seen during daylight.

The table shows some of the brighter comets that have illuminated our skies during the last 50 years. I hope you will find it as pleasant a nostalgia trip as I do...

Official designation	*Name*	*Orbital period (years)*
C/1956 R1	Arend-Roland	Infinite
C/1965 S1	Ikeya-Seki	880
C/1969 Y1	Bennett	1680
C/1975 V1	West	500,000
1P (1986 apparition)	Halley	76
C/1996 B2	Hyakutake	14,000
C/1995 O2	Hale-Bopp	2400
C/2002 C1	Ikeya-Zhang	341
C/2006 P1	McNaught	infinite

Source: Philip's Astronomy Encyclopaedia

PLANETARY PECULIARITIES: OUR IDIOSYNCRATIC SOLAR SYSTEM

Do the other planets have magnetic fields?

It's a great question, and one that has really only been answered in today's era of robotic exploration of the planets. Jupiter is the one exception to this mode of discovery. Its magnetic field is so strong that it reveals itself in the intense radiation belts that surround the planet and accordingly was detected by radio astronomers back in the 1950s. Jupiter's magnetic field is the most powerful of all the planets, some 20,000 times stronger than that of the Earth. At the other extreme is Venus, which has no detectable magnetism whatsoever. The rest of the planets lie somewhere in between.

As we saw in Chapter 3 ('Will the Earth's magnetic field reverse?'), our planet's magnetism arises because of the dynamo effect of its iron core, which has both solid and liquid components. This is probably the underlying mechanism for other planets' magnetic fields, too, although the details may be different. For example, one theory of Jupiter's internal structure is that it has a hydrogen core that has been compressed into electrically conducting metallic hydrogen by the enormous mass surrounding it. Perhaps the same applies to Saturn, but Uranus and Neptune are not massive enough for this to occur yet they, too, have magnetic fields stronger than Earth's. All four of the gas giant planets are known to display aurorae near their magnetic poles.

Mercury has a weak magnetic field, which is rather unexpected on account of its most undynamo-like rotation period (58.6 Earth days). But Mars' magnetic field appears to be merely a fossil, existing only in patches on the planet's surface, where the rocks are permanently magnetised. The Moon has similar magnetic properties. The suggestion is that both these worlds once had strong dynamo activity in their cores, which subsided as their interiors cooled. Not much point taking your compass there, then.

How do we know that some meteorites come from Mars?

At the time of writing, the tally of meteorites found on Earth that are known to have come from Mars stands at 34. Analysis of their stony composition shows that they fall into three groups, with the strange names of shergottites, nakhlites and chassignites. (Honestly, I'm not making this up.)

All three of these groups have chemical and isotopic (atomic structure) characteristics that do not occur on Earth, but are known to occur on Mars. The earliest identifications go back to the early 1980s, when Mars' atmosphere had recently been sampled by NASA's *Viking* spacecraft. Several other indicators have been used since then, all pointing consistently to a Martian origin of the meteorites. More recent analyses have allowed us to determine the ages of the Martian rock-beds from which they came, together with estimates of the length of time they spent wandering in space (typically 2–3 million years). It is even possible to investigate whether the rocks in which the meteorites originated had been exposed to liquid water on the planet's surface.

How did these objects get to Earth? It is assumed that they are the result of an asteroid or large meteorite hitting the Martian surface with such force that some of the fragments were projected into space faster than Mars' escape velocity, so that they became tiny planets of the Sun.

At least three such impact events are required to have produced the different groups of meteorites. After a few million years, the meteorites found themselves on a collision course with the Earth, eventually depositing themselves on the surface of our planet. Free gift samples of Martian rock are extremely rare, which is why these objects have been so well studied.

The most celebrated Martian meteorite is a potato-sized shergottite called ALH84001 (or more commonly, the Allan Hills Meteorite, from the part of Antarctica in which it was found in 1984). It contains tiny structures resembling terrestrial bacteria, which some scientists have interpreted as fossilised Martian life forms. The general consensus, however, is that this is extremely speculative and unlikely to be the case.

Why is Uranus upside down?

Actually, Uranus isn't really upside down (although the idea does provide a convenient way of remembering which of the planets we're talking about – U for Uranus). Unlike most of the planets, whose rotation axes stand more or less vertically in relation to the planes of their orbits, Uranus is tilted over on its side. Its axis points eight degrees *below* its orbit plane. Since that is nearer South than North, you could, at a pinch, describe the planet as being upside down (and also, therefore, as having retrograde rotation).

There is no known reason for Uranus' unusual stance in its orbit, but the best guess is a major collision event during the Solar System's turbulent infancy (see 'How did the Solar System originate?'). The energy required to induce such a large change in the axis of an object the size of Uranus would have been enormous, suggesting that the other object may have been a protoplanet of similar size to today's Earth.

What would we do if we found an asteroid on collision course with Earth?

Like all the best political responses, the answer to this question is 'It depends…'. But let's go back a few steps to set the question in context.

Back in 1979, two of my colleagues at the Royal Observatory in Edinburgh, Victor Clube and Bill Napier, produced cogent astronomical arguments suggesting that the Earth has suffered periodic impacts from asteroidal and cometary bodies, not only throughout its geological history, but in relatively recent times. These arguments were supported by the findings of an American physicist called Luis Alvarez, who found a layer of iridium-enhanced clay associated with the era in which the dinosaurs became extinct (the so-called Cretaceous-Tertiary boundary, of some 65 million years ago). Alvarez inferred that the extinction event was associated with one of Clube and Napier's impacts, and rather stole the limelight.

That theory of 'terrestrial catastrophism' is now generally accepted as being correct, and has led to the idea that the Earth is still at risk from collision with so-called Near-Earth Objects (NEOs) – asteroids and comet nuclei whose orbits intersect that of the Earth. Thanks principally to the efforts of NASA, which was nominated by the US Congress as having responsibility for coordinating the detection and scientific exploration of NEOs, a number of sky monitoring programmes have nearly achieved their goal of discovering 90 per cent of all NEOs bigger than 1 kilometre across. As noted in the introduction to this chapter, it seems certain that this goal will be extended to find successively smaller objects, since even a 250-metre impacting object could wreak nationwide devastation.

When such objects are found, the first step in evaluating their threat potential is to determine their orbits as far into the future as is feasible. To do this successfully requires a fairly lengthy series of observations. If it then turns out that there is, indeed, a serious risk of collision, other measures have to be introduced. The key issue here is time, and that's why the answer to the question has to be 'It depends...' If you have two or three decades' advance warning, then only a very slight change in the object's orbit will be enough to avert a disaster. Shorter timescales mean more drastic action is necessary.

Probably the first task to be undertaken would be to send a spacecraft to rendezvous with the rogue asteroid, and install a radio beacon so its exact trajectory could be monitored. Remember that these objects are in orbits not too dissimilar to Earth's, so they're not so far away. Then, given enough lead time, it may be possible to use rockets or explosives – some have even suggested solar sails – to give the object the slight sideways nudge necessary to alter its orbit. The shorter the lead time, the greater the impulse required, and no doubt *in extremis* nuclear weapons would be employed, despite the risk of one threatening object being transformed into many thousands of smaller ones.

The last few years have seen several space missions dedicated to improving our understanding of the structure of comets and asteroids. *NEAR-Shoemaker* landed on the asteroid Eros in February 2001, and *Hayabusa* sampled the asteroid Itokawa in September 2005. *Stardust* returned samples of Comet Wild 2 in

January 2006, while *Deep Impact* slammed a projectile into Comet Tempel 1 the previous July. In 2014, the ESA's *Rosetta* spacecraft will land a smaller probe on Comet Churyumov-Gerasimenko to study its chemistry. One day, perhaps, the information gleaned from such space probes may help to save our skins...

Do you think there is life elsewhere in the Solar System?

To be honest, yes, I do, and I think there's a good chance it will be discovered in my lifetime. The detailed answer to this question is very much the province of astrobiology – the study of life beyond the Earth – and we will revisit it in the next chapter. But at this point it's worth reiterating that so far, we know of nowhere else in the Universe where life exists other than here on our own planet. That means we have a sample of precisely one, which offers no clue as to how easily life can originate given the right conditions – or, indeed, how common it is.

The pragmatic view adopted by most astrobiologists, however, is that the Earth offers a salutary lesson in just how tenacious life is once it takes hold. For example, we now know of a whole class of bacteria called extremophiles – so named because they love extremes. These wild thrill-seekers can be found near boiling point around submarine hydrothermal vents, at freezing temperatures in Antarctic ice, and inside rocks. We know that some bacteria can even survive radiation bombardment in the vacuum of space.

We also know that life tends to favour liquid water, which is why so much of the robotic exploration of the Solar System in recent years has been a quest for indicators of water, whether they be classical erosion features in the dry landscape of Mars, dissociated hydrogen atoms detected above the Moon's south pole, or the icy geysers now known to emanate from Saturn's moon Enceladus. The possibility of liquid water underneath a thick ice mantle on Enceladus – and perhaps also on Jupiter's moons Ganymede and Europa – has mission planners scratching their heads for techniques to penetrate the said ice and sample the putative oceans beneath for living organisms. Likewise, planners of the next generation of Mars rovers are wondering how to test the water-ice permafrost now thought to lurk a few metres below the planet's surface.

There is a good chance that if we do find life elsewhere in the Solar System, it will turn out to have a common origin with living organisms on Earth. That in itself will be remarkable, showing that bacteria are able to migrate from one planet to another by some natural mechanism. But the discovery of extraterrestrial life will also be one of the most profound breakthroughs ever made by science – since we will then know for certain that we're not alone. In terms of its evolution, any life found in the Solar System is almost sure to be very primitive, but who can guess what benefits to humankind such a quantum leap in biodiversity might bring in its wake?

Of course, the more exciting prospect is the possibility of finding living creatures similar to ourselves. For that, however, we will almost certainly have to look far beyond the Solar System. If we want to find E.T., we will have to look to the stars.

CHAPTER 8

STARSTRUCK

OUR GALAXY FROM THE INSIDE

> 'Space is big. Really big. You just won't believe how vastly hugely mindbogglingly big it is. I mean, you may think it's a long way down the road to the chemist, but that's just peanuts to space. Listen...'

And that, ladies and gentlemen, is Chapter 8 in a nutshell – brought to you courtesy of *The Hitch Hiker's Guide to the Galaxy*. Douglas Adams' comic genius was nowhere more apparent than in his gift for understatement, and the opening paragraph of his apocryphal *Hitch Hiker's Guide* is a classic example. It's a great tragedy that Douglas is no longer with us. Our anguished planet can ill-afford to lose people with his whimsical view of life, the Universe and everything.

On the other hand, it's the job of thoroughly dispensable people like me to tell you just how big space really is. And, as it turns out, that's quite easy to do. It's just a question of stretching one's imagination a little – and you'll quickly discover it's not *that* much further than down the road to the chemist.

The starting point is the little star we call the Sun. Yes, it's big to us – roughly 100 times the diameter of the Earth – but in broader terms it is rather small, a dwarf star, in fact. At the other extreme are stars whose diameters are hundreds of times that of the Sun, stars so big that the Earth's orbit would fit inside them. Guess what they're called? Giant stars. Makes you wonder who thought up such clever names...

In fact, many giant stars are the so-called red giants we met in the last chapter (see 'When will the Sun stop burning?'), which are huge in extent, but contain little more material than the Sun does. Their rarefied outer layers have often been likened to a 'red-hot vacuum'.

What gives us the scale of space is not so much stars' sizes, however, but the distances between them. This is where things do begin to get a little mindboggling. Many people know that the nearest star to our Sun is an inconspicuous little object in the southern constellation of Centaurus, with the satisfyingly appropriate name of Proxima Centauri. Its distance from the Solar System is 4.22 light years, which means that the light we receive from it today began its journey 4 years 2 months and 19 days ago. But light travels at an incredible speed – 300,000 km/second (299,792.458 km/second, if you want to be exact) or, roughly, a billion kilometres per hour. So in four-and-a-bit years it covers some 40 million million kilometres. Does that mean anything to you? It certainly doesn't to me.

Now let's think of it like this. Suppose you could shrink the Sun to be the size of a small glass marble, about a centimetre in diameter, which you could hold in your hand. On the same scale, Proxima Centauri is another smaller marble – but it's

300 kilometres away. That is the distance from London to Liverpool, for example, or Sydney to Canberra, or New York to Boston. And remember, this is the distance to the *nearest* star. Suddenly, space does seem rather big.

What really brings home the distance to Proxima Centauri, though, is if you imagine yourself travelling there on the fastest spacecraft we have ever launched. That's a robotic machine called *New Horizons*, which is currently racing towards Pluto at the astonishing speed of 23 km/second. But shrink that speed to the same scale as our two marbles, and it turns out to be roughly the speed at which grass grows. The journey will take some 60,000 years to accomplish. Quite a trip. Incidentally, when I tell this story to schoolkids in my home town of Coonabarabran, they say 'Oh, that's really cool. What sort of grass would that be?' Country kids know that different types of grass grow at different rates – which I find rather inspiring, too.

Once you have come to terms with the fact that generally speaking, stars are a *very* long way apart, the rest is fairly straightforward. It's just more of the same, really. If you could zoom outwards from our Sun, you'd see increasing numbers of stars entering your expanding field of view – starting with Proxima Centauri, of course. By the time you were looking at a volume of space say 25 light years in diameter, there would be some 30 (mostly faint) stars visible, but beyond this the numbers would increase rapidly. Eventually, you would begin to see that the stars are not uniformly distributed, but often turn up in clusters or groups. And as you continued to zoom outwards from the Sun, you would start to see the structure of our home in the Universe – the Milky Way Galaxy.

Many people know that the faint, uneven band of light that runs all the way around the sky is the tell-tale sign that we live in a flattened disc of stars. It has been called the Milky Way since ancient times, although it was Galileo who, late in

1609, discovered that it is stars rather than milk that we see coagulating there, thus solving a long-debated philosophical problem. Few people actually see it at all these days. Most of us live under the light-polluted skies of big cities, which rob the Milky Way of its visibility.

Until the early years of the twentieth century, astronomers thought we lived near the centre of this disc of stars, but in 1919 its true size was fathomed by an American scientist called Harlow Shapley. Using distant star clusters as beacons, he was able to deduce that the Milky Way itself includes only the nearer stars of what is actually a gigantic system. Today we recognise that the stars we see in this faint band of light belong to the Sun's immediate neighbourhood in space – the so-called Orion Spiral Arm.

So what is this Milky Way Galaxy? We know from almost a century of study that if we could see it from the outside, it would have the appearance of a disc-shaped aggregation of stars, gas and dust, with prominent (and no doubt stunningly beautiful) spiral arms made of young stars. At the very centre of this giant Catherine Wheel lurks an exotic, super-massive black hole. There is other stuff too – vastly more than we can see – because 80 per cent of the Galaxy is made of dark matter, something we know to be present by virtue of its gravitational effect, but whose nature is still unknown. (There's more on dark matter in the next chapter.) Rather parochially, our Galaxy is distinguished from all the other galaxies in the Universe by the fact that its name starts with a capital letter. It is *the* Galaxy – although it's also often referred to as the Milky Way Galaxy.

Our Galaxy's proportions are truly staggering but, once again, the numbers themselves don't make much of an impact. The light that took 4.22 years to get to us from the nearest star, travelling at just over a billion kilometres per hour, would take 100,000 years to travel from one side of the Galaxy to the other.

And its roughly circular boundary encompasses something like 400 *billion* stars, each one a sibling of our Sun.

The Sun itself is just a minor entity in this celestial city. It is situated within the disk, roughly halfway between the centre and the edge, or about 25,000 light years from the centre. Fine – but how can we get our minds around the real size of the Galaxy? A clue to this came from a question I was asked a couple of years ago by a friend who is a science journalist – and who really should have known better. One of Sydney's newspapers had devoted an entire tabloid page to an artist's impression of the Galaxy, and it was adorned with a large arrow labelled 'SUN', pointing to the Sun's position. 'Why,' asked my friend, 'doesn't it say "EARTH"? Are they so close on this scale that they would be indistinguishable?'

The answer to that is, of course, yes – but it led me to do a calculation that blew me away. Imagine the picture of the Milky Way Galaxy again, but now, rather than being the size of a newspaper page, imagine it to be the size of the Earth itself. How far apart would the Sun and the Earth be on that scale? It's hardly believable, I know, but they would be separated by just 2 millimetres. I think that conveys with stunning eloquence the leap in scale when you go from the Solar System to the Galaxy.

One further heartwarming thing you might like to know is that the Sun participates with its neighbouring stars in the rotation of the Galaxy, so it is effectively in orbit around the galactic centre. Every 200 million years or so (a 'galactic year'), it completes a full circuit. To achieve this, the Sun is currently rattling along at about 250 km/second in the direction of the constellation Cygnus, the Swan. That's the Sun and the entire Solar System, lock, stock and barrel – planets, asteroids, comets and all. And we simply don't feel a thing...

It's rather fitting that this beautiful spiral galaxy lies in the constellation of Pavo, the Peacock. Unromantically named NGC 6744, it looks a lot like our Milky Way Galaxy would if we could see it from the outside. The gigantic disc of stars, gas and dust lies at a distance of 25 million light years from us, and is seen through a sprinkling of nearby stars belonging to our Galaxy. (Anglo-Australian Observatory/David Malin Images, imaged by S. Lee, C. Tinney and D. Malin.)

SEEING STARS: THE VIEW FROM EARTH

How many stars can you see with the unaided eye?

I bet it's nowhere near as many as you think. On a clear moonless night away from city lights, you can see about 3,000.

Of course, as soon as you turn a telescope or binoculars onto the sky you can see vastly more, because of the huge number of stars in our Galaxy. At the limit of naked-eye visibility, however, there are only some 3,000 individually identifiable stars above the horizon at any one time. If you don't believe me, have a shot at counting them.

How many stars are there in the Southern Cross?

If you look at the Australian flag, you will see five stars, and that's how many are easily visible in the constellation of the Southern Cross. Before we go any further, though, we should ask, 'What exactly do we mean by a constellation?' The dictionary definition is 'an arbitrary grouping of stars in the sky', but that hides a history that goes back at least 4,000 years for our modern western system of constellations. This is based on the star patterns of the ancient Greeks which, in turn, owe their origins to earlier Sumerian constellations. Many ancient peoples, including the indigenous people of Australia, identified their own star patterns and evolved legends associated with them, but these star groupings often look very unfamiliar. Perhaps the most foreign to our eyes are the 283 tiny constellations developed by the Chinese in the third century AD.

Modern astronomy defines 88 constellations, and their boundaries in the sky were fixed in 1928 by decree of the International Astronomical Union. Even though the positions of individual objects in the sky are exactly specified by coordinates analogous to latitude and longitude on Earth, the constellation in which a particular object lies still provides a useful guide to its general vicinity – in much the same way as the location of a town or village is broadly defined by the country in which it is to be found.

The Southern Cross is one of the brightest and most easily recognisable constellations in the entire sky, although it actually covers the smallest area. The long axis of the cross points directly to the south celestial pole, which lies four-and-a-half times the length of the cross away from its foot (see 'How are pictures of circular star trails made?' in Chapter 3). This little snippet of information is rather handy, since, unlike the northern sky, the southern sky doesn't have a bright pole star.

Constellations are formally known by their Latin names, and the Southern Cross is Crux Australis or, more often, just Crux. The importance of this nomenclature comes in the naming of individual stars, which are given a Greek letter signifying their place in order of brightness relative to the other stars of the constellation, followed by the constellation name in the genitive case (indicating possession). Thus, the brightest star in the Southern Cross is α Crucis, meaning 'alpha of Crux'. All very logical – at least, as a rule. Incidentally, α Crucis is the star at the foot of the cross.

Crux's five naked-eye stars take the first five letters of the Greek alphabet, but β (beta) Crucis (on the left-hand arm of the cross when it's seen 'upright') is also known as Mimosa – don't ask me why. This star has recently been discovered to have a faint companion star. That is a relatively common occurrence (see 'What are binary stars?'), but Mimosa's little friend is unusual in being faint in visible light but bright in X-rays.

Returning to the original question, the fact is that within the constellation boundary of Crux, telescopes reveal hundreds of thousands of stars – millions if the telescope is big enough. But I think the questioner really meant how many can you see with the naked eye – with a particular emphasis on the view from the nation's cities, where most Australians live. Light pollution reduces the contrast between the stars and the artificially brightened background sky to such an extent that ε (epsilon)

Crucis (the 'misfit' star to the lower right of the cross's centre) may disappear. And I noticed recently that from the terrace of the Sydney Opera House, even δ (delta) Crucis (at the right-hand end of the cross) is hard to see. I wonder how long it will be before sky watchers in Australia's cities see our national symbol not as a cross, but only as a triangle...?

Can you see some stars all year round?

Indeed you can, and another glance at the circular star trail picture in Chapter 3 will explain why. Stars close to the pole of the sky – North or South – simply circulate around the pole as the Earth rotates, never rising or setting. Such stars are known as 'circumpolar stars', and exactly which stars exhibit this phenomenon depends on the latitude of the observer. This is because, as we noted in Chapter 3, the height (in degrees) of the pole above the southern or northern horizon is equal to the latitude of the observer. Thus, for an observer at the equator, both poles lie on the horizon so no stars are circumpolar, whereas an observer at one or the other of the Earth's poles will see all visible stars as circumpolar.

There is another, quite different set of stars that can be seen for almost the whole year round, and they lie near the equator of the sky. When I first began visiting Australia in the late 1970s, I was astonished to discover that the constellation Orion – one of the most striking in the entire sky – falls into this rarely noted group. Orion is the Heavenly Hunter, not just in European tradition, but in many Aboriginal cultures, too. Perhaps that's because the familiar rectangle of bright stars, with the three closely spaced stars of the Hunter's belt at its centre, has more or less the same shape whichever way up you see it.

Orion stands on the celestial equator, which means that it can be seen from anywhere on the planet. In the northern hemisphere, the constellation is very prominent in winter, but disappears altogether from the night sky during the summer months, as it is too close to the Sun. Thanks to the 23.5 degree tilt of the Earth's axis, however – and the particular direction in which the axis is offset from the 'vertical' – Australians have the luxury of seeing Orion for almost the whole year. This is because in midwinter, when the Sun is closest to Orion in the sky, it is also at its furthest north, and therefore still beneath the horizon when the constellation is rising or setting.

Orion is at its best in the northern hemisphere during moonless winter evenings, when it shines brilliantly in the southern sky. To see the constellation in all its glory, you must get away from city lights. The most spectacular bit is the region known to astronomers as the Sword of Orion, which contains one of the brightest clouds of interstellar gas in the sky. The so-called Orion Nebula is a region of vigorous star formation located some 1,500 light years away, and is a magnificent sight even through binoculars, or a small telescope. But what we see with our visible-light instruments is merely the tip of the iceberg. Infrared ('redder than red') and millimetre-wave (radio) telescopes reveal the Orion Nebula to be a brightly glowing cavity in the side of a much larger dark cloud of gas and dust. The cavity is illuminated by four stars forming a compact group known as the Trapezium. They, too, show up clearly in a small telescope.

What's special about the star Sirius?

If you follow the line of Orion's Belt in either direction, you will come to another bright star some distance away from the Hunter. The stars you'll find in each direction are remarkably symmetrical about Orion. The one to the north-west is the reddish star Aldebaran, the eye of Taurus, the bull.

According to mythology, Orion has just dealt Taurus a hefty blow with his club, which is no doubt why his eye looks a tad red. Unfortunately, mythology doesn't record what happened next.

Following the line of Orion's Belt in the other direction, to the south-east, brings you to the brilliant white star Sirius – often called the Dog Star. This is the brightest star in the constellation of Canis Major, the Greater Dog (as distinct from Canis Minor, the Lesser Dog – and no, I'm not kidding). To astronomers, the Dog Star is rather more elegantly known as α Canis Majoris, and it turns out to be the brightest star (other than the Sun) in the entire sky.

What determines the brightness of a star as seen from the Earth is principally a combination of two things. One is the star's intrinsic brightness, which is related to its diameter and other physical characteristics, while the other is its distance. It's no surprise that distant stars look fainter than nearer ones, and that relationship is well defined by the laws of physics. A third factor influencing the apparent brightness of a star is the presence of dust along the line of sight, which can make it look further away than it really is. Astronomers have methods of allowing for this, however.

How do astronomers compare stars of different brightness? Since the time of Hipparchus, in the second century BC, they have used a curious, uneven measure called the magnitude scale, in which the brightest stars are said to be of the first magnitude, fainter ones of the second magnitude... and so on, down to the sixth magnitude, which denotes the faintest stars visible to unaided eyes. With the invention of the telescope, the scale was extended beyond the sixth magnitude.

The evolving science of photometry allowed this scale to be properly calibrated in the nineteenth century, and we now define a difference of one magnitude between one star and another as representing a change in brightness of 2.512 times. It's not quite as arbitrary as it sounds, because a first magnitude star is exactly 100 times brighter than one of the sixth magnitude, and 2.512 is the fifth root of 100 (try it on your calculator).

Once this calibration was established, it was recognised that some brilliant objects (including the brighter planets) have negative magnitudes, and Sirius is one such, with a magnitude of –1.44. The next brightest star is Canopus (α Carinae), also in the southern sky, with a magnitude of –0.62. Aldebaran, mentioned above, rocks in at fourteenth brightest in the sky. Its magnitude is 0.87.

Why is Sirius so bright? It results from a combination of a relatively high intrinsic luminosity (22 times that of the Sun) and a fairly modest distance of 8.6 light years – indeed it is the seventh-closest known star to the Sun. The Dog Star is also special (though not uncommon) in having a fainter companion star, formally called Sirius B, but affectionately referred to as The Pup. This object is a dense, very hot star at the end of its life, a type known as a white dwarf. Indeed, it was the first of its class to be identified.

Do the stars have colours?

Yes, they do, and even a cursory look at the night sky shows that while most stars are white in hue, a few are noticeably reddish or orange, and one or two (such as Sirius) have a bluish edge to them. The science of photometry (see previous answer) allows the brightness of stars to be measured accurately through a range of coloured glass filters, enabling astronomers to make very precise determinations of star colours. We now know that the colour of a star is related directly to its temperature, with cooler stars (less than 3,500°C) appearing reddish, and the hottest stars (up to 50,000°C) looking distinctly blue-white. The Sun is a yellow star with a surface temperature of 5,500°C.

Do all the stars we can see still exist today?

The listener who asked this decidedly intelligent question was familiar with the idea that when you look out into space, you're also looking back in time. Even when you look at the Moon, you're seeing it as it was in the past – albeit only 1.3 seconds in the past, because that's how long moonlight takes to make the trip to Earth.

Most of the individual stars we can see with the naked eye are within a few hundred light years of Earth, so the light-travel time, while long by human standards, is very short in comparison with the timescale on which stars evolve. Even the shortest-lived stars – wildly energetic hot giants – have lifetimes measured in tens of millions of years. So it is unlikely that any star visible with the naked eye will have done anything dramatic in the time its light has taken to reach us. There is, however, one extremely unstable star in our region of the Milky Way Galaxy, a colossal object with the name of η (eta) Carinae that weighs in at around 100 times the mass of the Sun. This star's erratic behaviour leads astronomers to believe it may be on the point of exploding as a supernova and, since it is in the region of 8,000 light years away from us, it might have already done so. But until the alarming pulse of radiation and subatomic particles reaches us, we simply won't know.

STAR PERFORMERS: HIGHLIGHTS OF THE STELLAR CABARET

What are binary stars?

As the name suggests, binary stars are ones that come in pairs. Surprisingly, more than half the stars in the sky are binaries of one sort or another – or are multiple systems with more than two components – so as a class they are pretty important. The essential ingredient of a binary system is that the two stars are connected by their mutual gravitational attraction. That is, they are in orbit around one another. To be technically correct, they both orbit their common centre of mass – or barycentre – whose position depends on the ratio of their masses. If they have equal mass, for example, the centre of mass will be halfway between them. This gravitational connection differentiates binary stars from what are called 'optical double' stars, in which two unrelated stars at different distances appear close together because they lie almost in the same direction in the sky.

The orbital period of binary stars – the time taken for each revolution about the centre of mass – varies enormously from system to system, ranging from hours to millions of years. This implies an enormous range of separation of the two components, with the shortest-period binaries being – literally – in contact with one another. Such extraordinary objects are called, not surprisingly, contact binaries, and they exchange mass and energy within a common envelope of material.

Less exotic binaries are far more commonplace, and are usually classified according to the way in which they are observed. Eclipsing binaries, for example, betray their dual personality by the fact that during each orbit one star passes in front of the other, dimming the light output from the pair. For this phenomenon to occur, the orbit plane must be seen edge-on from Earth, or eclipses will not take place. You will no doubt have spotted that this implies the system is too far away for the individual components to be seen separately, the binary nature of the star being revealed only by the way its light varies with time – its so-called 'light curve'. Once again, it is the science of astronomical photometry (see three questions ago) that allows this to be determined.

There is another type of binary star that takes its name from the technique used to identify its double nature. One of the most useful instruments at the astronomer's disposal is the spectrograph, which splits the light from stars and other celestial objects into its component rainbow colours, allowing a rich 'barcode' of information to be read from patterns of lines in the resulting spectrum. Those lines result from the presence of chemical elements in the stars' atmospheres, and each element has a unique spectral signature. Moreover, if the star is moving, a phenomenon known as the Doppler Effect shifts the lines in proportion to its line-of-sight speed, and this has a particular application with binary stars.

Imagine two stars in orbit around one another that are so distant from Earth that they look like a single star. Twice in each orbit, they will be moving sideways as seen from Earth, so the line-of-sight velocity for both stars will be zero, and their spectrum lines will be superimposed. But a quarter of an orbit later, one star will be moving towards the Earth, the other away. By virtue of the Doppler Effect, the spectrum lines of the two stars will have shifted slightly, in opposite directions. These two separate spectra reveal the binary nature of the star. Such systems are called spectroscopic binaries. This powerful technique can even reveal pairs in which one component is too dim for its spectrum to be seen, as the back-and-forth movement of lines from the other star betrays its presence.

Binary stars are important in astronomy and astrophysics because determining their orbital parameters can allow the masses of the two component stars to be measured exactly – independently of theoretical models. Thus they are often used as calibrators for projects involving measurements of other stars.

Do stars shake?

Indeed they do, and even our sedate and well-behaved Sun experiences tremors that tell us much about conditions in its interior. Because stars are basically huge balls of gas, they are subject to sound waves passing through their interiors, and these waves reveal themselves by slight oscillations of the stars' surface layers. Like vibrating strings in musical instruments, the stars oscillate in several different modes simultaneously, giving a characteristic 'sound' to the vibration. The detection and analysis of these oscillations is the province of an amazing branch of astronomy known as asteroseismology – the seismology of celestial objects.

Most of us are familiar with the idea of seismology on the Earth. Sensitive vibration recorders are used to record the passage of sound waves through our planet, and reveal what it means in terms of the underlying strata. It has applications from oil and gas exploration to the prediction of volcanic and earthquake activity, although the latter is still in its infancy. But how do you put a seismograph on a star – or even the Sun?

The answer is that you don't. Instead, you use the magical instrument described in the previous answer – the spectrograph. The Doppler Effect allows you to sense whether a star's surface is moving towards or away from you, and measure its speed. By plotting these velocities with time, the various modes of vibration of the star can be recorded and the results analysed to give detailed information about the star's interior. This, in turn, leads to accurate determinations of stellar parameters such as temperature, gravity and age.

In this work, the more precisely you can measure the velocities, the better established are the final results, and recent work by Tim Bedding and his team at the University of Sydney has explored stellar oscillations with astonishing accuracy. They have achieved a precision of 1 metre/second – a slow walking pace – and I'm delighted to say that much of their work has been done with the Anglo-Australian Telescope. Have a look at www.physics.usyd.edu.au/~bedding/animations/ to see movies of the various modes of oscillation in normal stars – dramatically exaggerated to show exactly what is happening.

The exploration of oscillations in the Sun is usually known as helioseismology, which differs from asteroseismology in that there is much more light available, and also that small areas of the solar surface can be isolated to get more accurate results. The ten million or so simultaneous modes of oscillation experienced by the Sun represent sound waves propagating at differing speeds and depths, and typically have periods of around five minutes. The Global Oscillation Network Group (GONG) of solar telescopes around the world monitors the Sun continuously, to yield remarkably accurate investigations of our star's internal structure.

There is another type of 'shaking' in stars, and this is the large-scale pulsations that most stars experience at a late stage in their lives. Compared with the oscillations of normal stars measured by asteroseismologists, these pulsations are *big* – up to 30 per cent of the star's diameter. They are caused by large-scale instabilities in the stars' atmospheres, and occur with periods ranging from hours to years. The most obvious manifestation of this pulsation is a regular change in the star's brightness, which can once again be investigated using the science of photometry. Such 'pulsating variable stars' occur in various categories (depending on the star's mass), and have long been extremely useful in providing 'standard candles' for the exploration of our Galaxy and the wider Universe.

Could brown dwarf stars have planets?

Hang on a minute. What's a brown dwarf star? In the last chapter, we found that stars are formed by the collapse of a cloud of hydrogen gas under gravity, leading to an increase in pressure and, hence, temperature – and ultimately to the onset of nuclear reactions. Whereupon the protostar begins shining and becomes, well... a star. But what if the collapsing gas cloud is not massive enough to allow sufficiently high temperatures to be achieved? In that case, the object may become a brown dwarf star, a kind of intermediate stage between a gas giant planet like Jupiter and a regular hydrogen-burning star. Unlike normal stars, which emit their energy over a range of wavebands (including visible light), brown dwarfs radiate their meagre output in the infrared region of the spectrum.

Brown dwarfs typically have masses between 1 and 8 per cent of the Sun's mass and, in fact, are technically defined by the International Astronomical Union to have 'true masses above the limiting mass for thermonuclear fusion of deuterium (currently calculated to be 13 Jupiter masses for objects of solar metallicity)'. Make of that what you will. But brown dwarfs are important in as much as these small, dim objects probably exist in very large numbers in our Galaxy.

So, could brown dwarfs have planets orbiting around them? Yes, they could – and they probably do. At least one of the recent campaigns to search for extrasolar planets (planets orbiting stars other than the Sun) concentrated specifically on brown dwarf stars, and for a very good reason. A serious problem that arises in trying to obtain direct images of extrasolar planets is that the planet is *much* fainter than the star. In the case of a normal star such as the Sun with an Earth-like planet orbiting around it, the planet is some ten billion times fainter. At the distances such objects are observed, the planet appears to nestle close to the star, and is simply lost in its glare.

But now imagine a planet in orbit around a dim brown dwarf star. The star would not be all that much larger than the planet you're looking for, so the difference in brightness is very much less. That consideration has prompted the direct search for extrasolar planets mentioned above, and at least one promising candidate has been found. No doubt we will hear more as this research progresses.

What is the average distance between the stars?

If I've done my calculations correctly, the density of stars in the Sun's immediate neighbourhood (within 12 or so light years) is 0.004 stars per cubic light year (or one star per 250 cubic light years), suggesting an average separation between any pair of stars of around eight light years. That figure is skewed by the fact that almost half the 29 stars within this volume are members of pairs or multiple systems, so the typical separation between groups of stars is rather more than that. It means that the Sun's closest neighbour, Proxima Centauri (4.22 light years' distant, and a member of the α Centauri triple system), is closer than average.

It's also worth remembering that the Sun lies within the disc of our Milky Way Galaxy, where the star density is relatively high. In the space between galaxies the star density drops to zero, while at the centre of a globular cluster (a compact aggregation of hundreds of thousands or millions of stars) it may exceed five stars per cubic light year.

Are there molecules in space?

Molecules are combinations of atoms, and you might think they belong in the province of chemistry rather than physics and astrophysics. Indeed, most chemists would regard the environment of space as being totally hostile to the existence of molecules, because the high levels of ultraviolet radiation from stars would have the effect of dissociating the molecules into their individual atoms. But astrochemistry is a topic that has assumed enormous importance in recent years, and may even give us significant clues to the origin of life itself.

Some molecules, such as titanium oxide (TiO) and cyanogen (CN) are sufficiently robust that they can survive in the atmospheres of cool stars, and are prominent in those stars' rainbow spectra. But for more fragile molecules to survive, they need to be shaded from radiation, and it turns out that they often are – by themselves. This is because they collect in clouds, which are sometimes extremely large – in excess of 10,000 times the mass of the Sun (at which point they become officially known as 'giant molecular clouds', or GMCs). The molecules exist in either gaseous or solid forms in space, the solid particles comprising what we refer to as interstellar dust.

The detailed exploration of molecules in space requires the use of sensitive radio telescopes, and it was only in 1963 that the first interstellar molecule was discovered by this means. That was the hydroxyl molecule (OH), which was quickly followed by carbon monoxide (CO), the second most common molecule in space after molecular hydrogen (H_2). Water vapour (H_2O) and ammonia (NH_3) were found soon afterwards. The discovery of alcohol (C_2H_5OH) naturally raised excitement levels among party-going astronomers, but the recent detection of amino acids – the building blocks of terrestrial life – has prompted the suggestion that perhaps we owe our very existence to interstellar chemistry. There are now well over 100 different molecules known in space, and it seems certain that this field of discovery will continue to expand.

EXOTICA GALACTICA: TOWARDS THE FRONTIERS OF KNOWLEDGE

What are neutron stars made of? And what are pulsars?

Yes, I wondered when we'd get on to this. If I may, though, I'd like to go back a couple of steps first. In the last chapter ('When will the Sun stop burning?'), we discovered that the collapsed core of a star such as the Sun eventually turns into a compact object called a white dwarf star. We see many such stars in the Sun's vicinity, mainly because becoming a white dwarf is the ultimate fate of more than 95 per cent of the stars in our Galaxy. Commonplace though they are, these elderly objects have very unusual properties. Something with a mass comparable to that of the Sun is compressed into a volume about the same as that of our planet. That results in an average density measured not in grams per cubic centimetre, as with most Earthly substances, but tonnes per cubic centimetre – or even *tens* of tonnes per cubic centimetre.

This suggests that the material from which white dwarfs are made is in a very unusual state, and indeed it is. It's described, rather unkindly, as degenerate matter or, more specifically, electron degenerate matter. What that means is that electrons no longer orbit the atoms' nuclei in the normal way, but are squeezed together alongside them, so that only the natural mutual repulsion of the electrons stops the material from being compressed any further under its own gravity. White dwarfs begin as extremely hot objects, but it is thought that over tens of billions of years, they will cool to become frozen, inert black dwarf stars.

Back in 1931, the great Indian-American astrophysicist Subrahmanyan Chandrasekhar discovered that a white dwarf with a mass of more than about 1.4 times that of the Sun (the so-called Chandrasekhar Limit) can't stay as a white dwarf. Its extra gravity will overcome the pressure of the electrons, leading to further collapse until only the degeneracy pressure of the neutrons in the atomic nuclei will hold things apart. And – you've guessed it – that's what neutron stars are made of.

Neutron stars represent an even more unusual state of matter than white dwarfs, with densities in the region of a billion tonnes per cubic centimetre. In size, a typical neutron star is only 10–20 kilometres across, but once again it is of comparable mass to the Sun. Unlike white dwarfs, neutron stars are produced in a matter of hours when stars whose initial mass is around ten times that of the Sun detonate in violent supernova explosions. Incidentally, this dramatic process, involving nuclear reactions at extremely high temperatures, produces the heavier species of atoms. So if you're wearing a gold ring as you read this, the atomic nuclei from which your ring is made began their lives in a colossal explosion in space. And how cool is that?

Finally, how does a neutron star become a pulsar? The term originated in the 1967 discovery by Jocelyn Bell and Antony Hewish at Cambridge of rapidly pulsating radio objects whose nature was quite unknown. 'Little green men' was one early hypothesis. But it turns out that the high density and incredible strength of neutron stars allows them to rotate rapidly, while their concentrated magnetic fields cause them to beam radio waves out from their magnetic poles. If the geometry is such that the Earth is in line with this beam as it is swept, lighthouse-like, by the neutron star's rotation, then you've got a pulsar.

Well over 1,000 of these objects are known, most with pulse periods of around one second, although some (known as millisecond pulsars) have much shorter periods. They form extremely accurate clocks, with a well-understood rate of slow-down due to the braking effect of the strong magnetic field. That super-precise timing has led to the discovery of planets orbiting one pulsar (with the improbably sexy name of PSR B1257+12) and, in an unusual binary pulsar, the confirmation of the existence of elusive gravitational waves, as predicted by Einstein's General Theory of Relativity (see 'Does gravity have a speed?' in Chapter 10).

Where is the nearest black hole?

I don't know the answer to this question. In fact no one does, since we can only observe black holes by their effect on their surroundings, and it's possible there may be many that remain undetected. Let me say a bit more about the topic, though, beginning with a few words about the nature of a black hole.

If you've read the previous answer, you'll know that some massive stars collapse at the end of their lives into exotic neutron stars, in which only the mutual repulsion of subatomic neutrons prevents further collapse. If the star is even more massive – well in excess of ten times the Sun's mass – then the neutrons are unable to stop the rot. The core of the star will just keep on collapsing to produce a singularity – a point where the density of space becomes infinite. Although the mathematics of such singularities were developed soon after Einstein published his General Theory of Relativity in 1915, it wasn't until 1967 that John Wheeler coined the term 'black hole'. Such objects are famed for having an 'event horizon' beyond which nothing – not even light itself – can escape (hence the name), and for 'spaghettifying' any unfortunate victims who stray too close, due to the difference in gravity between their heads and their feet. *RIP*.

Since we can't see black holes, we have to look for things they can do to their surroundings to be able to detect them. The first black hole candidate, discovered in the 1960s, was a bright source of X-rays called Cygnus X-1, which is a binary star in which one component is invisible, but calculated to have about ten times the mass of the Sun. That's the black hole. It is thought that the X-rays come from extremely hot gas in a rapidly rotating disc of material surrounding (and being swallowed by) the black hole. Cygnus X-1 is about 6,000 light years away from us, and X-ray observations have allowed many other similar candidates to be discovered since.

Apart from these so-called 'stellar mass' black holes, there are two other generally accepted categories. One is entirely theoretical, and comes from the possibility that large numbers of mini-black holes might have been created in the aftermath of the Big Bang. Such primordial black holes may, in fact, have evaporated by now, as a phenomenon known as 'Hawking radiation' eventually strips black holes of their mass, and the lower the mass of the object, the quicker this happens. Yes, Hawking radiation was predicted theoretically by Stephen Hawking – as were the primordial black holes themselves.

The third category is the so-called 'super-massive black hole', which seems to occur commonly in the centres of galaxies like our own. These have masses measured in millions of times that of the Sun, but the black holes lurking in the centres of so-called active galaxies and quasars (see 'What's a quasar?' in Chapter 9) may have masses hundreds of times bigger still. We have very good evidence that there is a black hole with a mass of 3.6 million times that of the Sun at the centre of our own Galaxy, some 25,000 light years from the Solar System. The accurate determination of its mass comes from the fact that we can see stars orbiting around it – a bizarre phenomenon, since the black hole itself doesn't show up at the infrared wavelengths with which we can see the stars, so it looks for all the world as if they are orbiting around... nothing. However, the characteristics of that black hole are now well determined, and it is certainly the nearest super-massive black hole to the Sun.

Finally, many astronomers have wondered why there seem to be black holes of similar mass to the Sun, and black holes weighing millions of times the mass of the Sun – but nothing in between. This is still a puzzle, but some very recent observations suggest that there may be 'intermediate mass' black holes lurking at the centres of some globular clusters – ancient aggregations of hundreds of thousands of stars that accompany many galaxies. If this is confirmed, it will give the theoretical astronomers trying to build models of intermediate mass black holes something meaty to work on...

Could the Solar System have a twin anywhere?

While the existence of an exact identical twin is rather unlikely, it seems impossible to imagine that among the 400 billion stars in our Galaxy, there wouldn't be one with a planetary system that closely resembles our own. If you extend the 'anywhere' to include the other 100 billion or so galaxies in the Universe, then an exact match becomes an even greater possibility.

The reason we can say this with some confidence is that since 1995, when the first one was discovered, we have known that other stars do have planets. Before that, the idea was a theoretical likelihood rather than a proven fact. In the last few years, however, there has been a massive effort by astronomers all over the world to find more of these so-called extrasolar planets, and as of mid-2007 the total is about 250.

There are several different techniques used to find such objects, almost all of which require careful analysis of the light from the parent star rather than direct observation of the planet. The reason for this is the minuscule angular separation of the planet and star as seen from Earth, combined with the dazzling brilliance of the star itself (see 'Could brown dwarf stars have planets?'). The most successful technique is the one employed in the Anglo-Australian Planet Search programme, which uses the 3.9-metre Anglo-Australian Telescope at Coonabarabran in north-western New South Wales. This so-called 'Doppler wobble' technique relies on the fact that an invisible planet orbiting a distant star imparts a slight reflex motion to the star, since the two objects actually revolve around a common centre of mass rather than the centre of the star itself.

In an exactly analogous method to the analysis of spectroscopic binary stars (see 'What are binary stars?'), it is possible to use a spectrograph to detect the back-and-forth movement of the star's spectrum lines as the planet progresses around its orbit. Because planets are much less massive than stars, however, the change of speed induced in the parent star is tiny. For example, Jupiter imparts a back-and-forth motion on the Sun amounting to only 12.5 metres/second. When you consider that astronomers are more used to measuring velocities in kilometres/ second and, moreover, that this minute change of speed occurs slowly over six years or so (since Jupiter's orbital period is 11.86 years), you get some idea of the problem.

At the Anglo-Australian Observatory, star velocities can now be measured with an accuracy of about 1 metre/second, so Jupiter would easily be discovered if aliens were using the same equipment to observe the Solar System. Earth-like planets are at present below the level of detectability, however – although so-called 'super-Earths', with masses of 10 to 20 times that of our planet, probably aren't.

To date, the various extrasolar planet search programmes have turned up a large number of 'hot Jupiters' (planets of Jupiter's mass or greater) skimming close to the surfaces of their parent stars and often in elongated orbits. Several multi-planet systems are known, the intertwining periodicities of the individual planets having been disentangled by a sophisticated mathematical process called Fourier deconvolution. While the Holy Grail of this work remains the discovery of planetary systems like our own, we can take heart from the fact that at least 10 per cent of the stars in the Sun's neighbourhood seem to have planets around them – and perhaps many more.

Do you think there is intelligent life out in space?

A number of the answers in this book have touched on issues to do with the evolution of life in the Universe, for example in the Moon's role in stabilising the Earth's rotation (Chapter 6) and the possibility of bacterial life existing elsewhere in the Solar System (Chapter 7). The fact remains, however, that we know of nowhere else in the Universe where life exists other than Earth. The relatively new discipline of astrobiology brings together the various fields of study that have some bearing on this problem. Thus biology, astronomy and planetary science are key components, but other relevant disciplines such as chemistry, geophysics, palaeontology and climatology are also involved.

If the Holy Grail of planet-finders is another Solar System, then the Holy Grail of astrobiologists is intelligent life. And it seems certain that in order to find it, we will need to look to the planets of other stars. We saw in the introduction to this chapter that with present technology, a journey to the nearest star would take tens of thousands of years. Thus, our only recourse is to remote sensing – to exploring these planets with large telescopes.

One of the drivers behind the push to build steadily larger telescopes is the possibility of capturing actual images of extrasolar planets rather than merely inferring their presence from indirect observation. This will require telescopes twice the size of today's largest, with mirrors at least 20 metres in diameter. What these monsters will do is to disentangle the faint image of a planet from the glare of its parent star, allowing a detailed analysis of its light using the spectrograph. As noted in Chapter 6 ('What is earthshine?'), this will allow the detection of any life-sustaining compounds in the planet's atmosphere. If so-called biomarkers are found, living organisms could be present.

But to establish the presence of *intelligent* life, we would need to find unequivocal signs of intelligent activity. One possibility would be the existence of industrial pollutants such as carbon dioxide and heavier chemical compounds in the planet's atmosphere. Such a find would be truly astonishing, and would profoundly affect the way we see ourselves and our place in the Universe. Even the fact that the newly discovered life-forms were trashing their planet with similar recklessness to the way we treat our own would do little to diminish the excitement of such a discovery.

Many people are familiar with another proposed method of detecting intelligent life in space, and that is the Search for Extraterrestrial Intelligence, or SETI. For almost half a century, this well-directed series of experiments has been conducted – mostly using large radio telescopes – to look for signals that could be interpreted as signs of intelligent communication by alien species. To date, nothing has been found.

One of the early pioneers of SETI was Frank Drake, author of the famous Drake Equation. This equation combines the various requirements for intelligent life to evolve into an estimate of the probability of intelligent species inhabiting our Galaxy. In Drake's time, all the parameters in the equation were guesses. Now, however, we do at least know one of them – that, yes, stars commonly have planets around them. No doubt in time, others will follow.

Do I think there is intelligent life out in space? To be honest, I find it hard to believe that there isn't. But I'm not laying any bets as to when we'll find out.

AND FINALLY...

How do you measure light years?

This is a great question, and I'm sure it's one that conjures up images of astronomers with stop-watches timing beams of light as they fly through space. Of course, that's impossible. So how do we measure light years? Well, we don't. In fact – and I hope this won't come as too much of a shock – in the trade, we don't even *use* light years.

A light year, as we discovered in the introduction to this chapter, is a unit of distance. It's the distance travelled by a beam of light (in a vacuum) in one year, and amounts to 9.46 million million kilometres. It's a hell of a long way. The unit of distance used by astronomers, however, is one that can be measured directly – unlike a light year – and it's called a parsec. It's not often astronomers give sensible names to things, but this is one they got right, for at a distance of a parsec, a star or other celestial object has a *par*allax of one *sec*ond of arc.

What does that mean? Geometry tells us that one second of arc (or arcsecond) is a tiny angle amounting to 1/3,600th of a degree. So much for geometry – it's much more instructive to imagine a person 5 kilometres away holding up a coin. A British pound, an Australian dollar and an American quarter are all about the right size. To your eye, the coin's diameter is one second of arc – and you would need a sizeable telescope to be able to see it. Now imagine the Earth moving around the Sun in its orbit. As the Earth's position changes, the angle to your star or whatever changes, too, and this is what is meant by parallax. If the object is one parsec away, that angular change is one second of arc. *Voilà!*

To be exact, a parsec is the distance at which the radius of the Earth's orbit (not its diameter) makes an angle of one second of arc. The first parallax of a star was measured by Friedrich Bessel at Königsberg in 1838, and the process is now completely routine, using automated equipment carried aboard purpose-built satellites. It only works for stars out to a few thousand parsecs or so, however, and beyond that other means of estimating distances have to be used.

As it turns out, parsecs and light years are not all that different in scale. A parsec is 3.262 light years. So why do popular astronomy writers always quote distances converted into light years, rather than stick with the original units? I think it's partly because they are easier to understand but, perhaps more importantly, they convey with far greater eloquence the utter vastness and emptiness of space.

CHAPTER 9

ACROSS THE UNIVERSE

THE REALM OF GALAXIES

What do we see when we look beyond our Galaxy? We have to do this away from the band of the Milky Way, because the stars there are accompanied by clouds of dust, which block our view through the thickness of the disc. But if we look upwards or downwards with respect to the plane of our Galaxy we see... other galaxies. Lots of them.

In fact, our Galaxy is one of the two largest members of a small cluster of about 40 galaxies called the Local Group (parochialism again coming to the aid of the nomenclature). This little galactic shire encompasses a span of perhaps five million light years, and includes many dwarf galaxies, as well as two small galaxies known as the Magellanic Clouds. These fuzzy objects are prominent in the southern hemisphere sky, and look for all the world like bits of the Milky Way that have somehow broken off.

The other large member of the Local Group is the famous Andromeda Galaxy, which has the prestigious honour of being the Most Distant Object Visible to the Unaided Eye. The light that enters your eyes during November evenings when Andromeda is at its best has been on its way for some 2.5 million years – much longer than *Homo sapiens* has walked the Earth. One of the things that astronomers are able to measure fairly easily is the relative speeds of our Galaxy and the Andromeda Galaxy, and it turns out that they are approaching one another at about 100 km/second. We're expecting a collision in about three billion years, so hold on to your hats.

Beyond the Local Group, galaxies dominate the sky. We find great clusters with hundreds of member galaxies, such as those seen in the constellations of Virgo and Coma Berenices (Berenice's Hair). Though they themselves are undoubtedly spectacular, such clusters also group together to form even bigger aggregations of galaxies. Not surprisingly, they are known as superclusters and the Virgo cluster is at the core of our Local Supercluster – in which the parochialism of the nomenclature truly outdoes itself.

Clusters and superclusters represent the largest-known concentrations of matter in the Universe and, just like the Earth, they exert a gravitational pull on their surroundings. It can be shown, for example, that the enormous gravitational attraction of the Virgo Cluster is drawing the Local Group towards its centre. This observation is complicated by the fact that the Virgo Cluster is at present being carried *away* from us by the expansion of the Universe (of which more in due course), so it will take time for Virgo's gravity to overcome that and pull us in. Here, I'm reminded of some advice from the prominent American amateur astronomer, John Dobson. He once suggested that the best way to imagine the Universe is not to think of ourselves looking up into the sky, but looking down into it. In those terms, it's very easy to imagine ourselves plummeting into the Virgo Cluster.

Virgo's distance of 65 million light years means that there is no immediate cause for alarm. The time to impact is measured in tens of billions of years. In any case, galaxies and galaxy clusters contain so much empty space that 'impact' is completely the wrong expression – 'tidal disruption' is a much better one. So there's probably no need to hold on to your hat after all. And, in the meantime, here are some questions that radio listeners have posed in relation to our intergalactic surroundings and the wider Universe.

SPACED OUT: GALAXIES BY THE BUCKETFUL

Is it true that there are more stars than there are grains of sand on all the Earth's beaches?

The best thing about this question is that it makes you get your head around some *very* big numbers. I think the original quotation is due to Carl Sagan, one of the greatest astronomy popularisers of all time, and came from his *Cosmos* TV series.

Although I always relish putting lots of zeroes after big numbers, the only sensible way to make this comparison is to use that familiar mathematical notation in which powers of ten are used. Thus, 100 is 10^2, and 27 million is 27×10^6. As it turns out, the easier part of the process is to calculate the number of stars. We saw in the last chapter that the best estimate of the number of stars in our Galaxy is 400 billion, but ours is actually quite a big galaxy (as is only fitting), so let's say that a typical galaxy will contain 100 billion, or 10^{11} stars.

Now, how many galaxies are there in the Universe? Observations made with the Hubble Space Telescope suggest that the instrument is capable of detecting some 80 billion galaxies, and most astronomers would accept that this is a fairly conservative estimate. The total number of galaxies is probably at least 100 billion, or 10^{11} once again. Combining this figure with the number of stars in a galaxy gives an estimate of the total number of stars in the Universe – roughly 10^{22}.

The trickier bit comes in estimating the total number of grains of sand on all the Earth's beaches, as it depends on how big you think a grain of sand is. Taking an average size of 0.5 millimetres, and combining it with some reasonable assumptions about the average width and depth of beaches – plus the fact that the Earth has roughly one million kilometres of coastline (of which, apparently, some 36 per cent is sandy) – gives an estimate of the total number of sand grains as... wait for it... 10^{21}. So yes, based on those figures, there are indeed more stars than grains of sand, and Carl Sagan was right.

Two things that interest me about this are first that the two numbers are not terribly dissimilar, meaning that Sagan made a very good choice in picking grains of sand for his comparison with the number of stars. But also, we know from other studies that stars make up only about 0.5 per cent of the total contents of the Universe. In the general scheme of things, they are not very important at all.

So what is the rest made of? Very roughly, it's another half per cent of subatomic neutrinos whizzing around the place, about 4 per cent of hydrogen and helium left over from the Big Bang, and then a whopping 20 per cent of something called 'dark matter'. Finally, 75 per cent of the Universe consists of 'dark energy'. These two dark secrets constitute the biggest mysteries facing astronomers today, because we have precious little idea of what they are. They are discussed further in this chapter and the next. But where is the familiar stuff from which planets, mountains, cities and people are made in all this – oxygen, silicon, carbon, nitrogen, iron and all the rest?

The answer is that together, they amount to about 0.03 per cent. Makes you feel pretty small, doesn't it?

With millions of galaxies in the Universe, how does the light get through?

The answer to this question follows directly from the end of the last one – the Universe is mostly made of things that we can see straight through. Space itself is completely transparent, as are the biggest constituent components of the Universe's mass-energy budget – dark matter and dark energy. Effectively, therefore, the Universe looks as if it is mostly empty space, with objects such as galaxies making up a very small fraction of the total contents. This means that even though light rays might travel through space for many millions or even billions of years, the Universe is so devoid of solid matter that they don't bump into anything.

This question is also related to something known as Olbers' Paradox, named after a nineteenth-century German amateur astronomer called Wilhelm Olbers – although he was by no means the first person to think of it. The paradox is that if the Universe is infinite and filled with stars, then in whatever direction we look we should see a star. So why is the sky dark at night? Today, the resolution of this paradox is usually given in terms of the overall expansion of the Universe, which stretches the wavelength of radiation moving through it. Thus the light from the most distant objects has been stretched, or 'redshifted', out of the visible region of the spectrum, beyond the red and into the infrared. As we shall see in the next chapter, this is just as well, or the sky would still be filled with the blinding afterglow of the Big Bang itself...

How do we know there is dark matter?

The American amateur astronomer John Dobson once described the Universe as 'mostly hydrogen and ignorance'. However, he was wrong. As we saw two questions back, hydrogen only amounts to some 4 per cent of the Universe's mass-energy budget. In reality, the Universe consists mostly of dark matter and dark energy – and ignorance. And anyone accustomed to hanging out with astronomers will know that they feel acutely embarrassed at not knowing what these two dark mysteries are.

In some ways, dark matter is the easier of the two to get your head around. It reveals itself by the effect of its gravitational pull on matter that we *can* see. Other than that, however, we have no way of detecting it, although we can investigate its characteristics through a number of techniques.

While many people imagine that the quest to understand dark matter is a recent development, it was back in 1933 that the American-Swiss astronomer Fritz Zwicky first noticed that something didn't add up. He was investigating clusters of galaxies – the largest concentrations of matter in the Universe – and was particularly interested in a very rich cluster of galaxies in the constellation of Coma Berenices.

The spectroscopic technique that enables star velocities along the line of sight to be measured accurately (see 'What are binary stars?' in Chapter 8) works equally well for galaxies. Zwicky used it to determine the speeds of several members of the cluster. He was amazed to find that these galaxies seemed to be moving too fast for the cluster to hold on to them. Using a mathematical device known as the virial theorem, he established that the gravitational attraction of the cluster's visible matter – the stars, gas and dust in the constituent galaxies – was not enough to bind the cluster together.

Given their velocities, the galaxies he was observing should have escaped from the cluster long ago. Zwicky inferred from this the presence of something else, an invisible component that neither emitted light nor absorbed it from the radiation of background objects. Whatever this component was, it exerted a gravitational pull on the members of the cluster. He was right on the money with this deduction, but little attention was paid to it at the time.

Forty-five years later, the great American astronomer Vera Rubin (a heroine of mine – and not just because she wrote a nice review of my last book) made accurate measurements of the rotational velocities of galaxies, and discovered new evidence for the existence of dark matter. We know that individual galaxies rotate, and by selecting disc-like spiral galaxies that are tilted almost edge-on to us, we can measure the way the rotational speed changes with the distance from the centre of the galaxy. Such velocity profiles are known as rotation curves.

If the matter in a given galaxy is concentrated where the light of its stars is concentrated – as you'd expect – the rotation curve should show stars close to the centre of the galaxy moving rapidly, while those further away are moving more slowly – just as satellites behave in Earth orbit (see 'How do satellites stay up in space?' in Chapter 5). But that was not what Vera Rubin's measurements showed. Far from falling off with distance from the galaxy's centre, the rotation curve stayed almost constant out to its extremities. Rubin's observations were made with clouds of gas rather than stars, but the bottom line was the same. These rotation curves could only be explained if the galaxies were enveloped in gigantic spherical halos of so-called dark matter.

Since then, several different types of study have demonstrated the existence of dark matter, and given us some insight into its nature. The three main lines of investigation are:

1. *Gravitational lensing* One of the consequences of Einstein's General Theory of Relativity (still the best theory of gravity we have) is that matter distorts space. We feel that subtle bending as gravity. Rays of light feel it, too, and as they go past a massive object such as a galaxy or cluster of galaxies, they are deflected slightly in a process known as gravitational lensing – because the deflecting object behaves rather like a glass lens. But the lens effect comes from the action of both visible *and* dark matter, so it can be used to investigate the distribution of both types of material in the object. Gravitational lensing has shown us that galaxies are embedded in large volumes of dark matter, exactly as suggested by Vera Rubin's measurements.
2. *Galaxy surveys* By measuring the distances to many tens of thousands of galaxies, it is possible to show how visible matter is distributed in the Universe on very large scales. The Anglo-Australian Observatory 2dF Galaxy Redshift Survey pioneered such studies in the late 1990s with a survey of 221,000 galaxies out to about 2.5 billion light years. (The reference to 2dF (2-degree field) comes from the name of the instrument on the 3.9-metre Anglo-Australian Telescope that was used to make the survey: a spectrograph with a field of view on the sky of 2 degrees, able to observe 400 galaxies at a time.) The survey showed first and foremost that the distribution of galaxies in the Universe is far from uniform, with galaxies concentrating along the edges of great voids in space with a characteristic size of around 300 million light years – almost like a foam of galaxies. But because all kinds of matter – visible and dark – affect the geometry

of space (by the distortion predicted by general relativity), we can investigate the relative distribution of the two types statistically. This tells us that the ratio of dark matter to visible matter in the Universe is about four to one, and that the two types of matter concentrate together. 'Beacons on the hills of dark matter' is an apt description of galaxies by Matthew Colless, who led the 2dF Galaxy Redshift Survey.

3. *Star velocity surveys* Knowing that dark matter concentrates wherever visible matter is found suggests another approach to its investigation. Our own Milky Way Galaxy is a giant spiral system of 400 billion stars with associated gas, dust and dark matter. So dark matter is all around us. How is it distributed? Does it occur in clumps and, if so, how big are they? What might we learn from their size? These questions can be investigated by measuring the velocities of hundreds of thousands of stars to look for streams of stars in particular groups, revealed by their common speed. Since the stars are moving under gravity – which comes from both visible and dark matter – we can use the stars' motion to map the underlying gravity field and identify the dark matter component. One outcome of this is an estimate of the minimum size of a clump of dark matter. Measurements suggest it may be around 1,000 light years across. If that is the case, we can deduce from thermodynamical arguments that its temperature is a few thousand degrees above absolute zero, rather than the few tenths of a degree that is usually assumed.

So what is dark matter? The best guess today is that it consists of some sort of massive subatomic particle that exists in huge numbers, but interacts only weakly with normal matter (what's known as a WIMP – or weakly interacting massive particle). The competing theory of MACHOs (massive compact halo objects – such as brown dwarf stars) now seems less likely.

One other aspect of dark matter is its vital role in the evolution of galaxies. Theoretical models suggest that in the early Universe, dark matter clumped together, and the gravity of the clumps caused hydrogen left over from the Big Bang to concentrate into clouds – which eventually collapsed into stars and galaxies. This leads us to infer that dark matter seen at greater distances (that is, at greater look-back times) should be less clumped than dark matter seen close by (in the more recent past). Gravitational lensing studies with the Hubble Space Telescope already indicate that this is, indeed, the case.

If Newton's law of gravitation varied, would it eliminate the need for dark matter?

The listener who asked this question was clearly well in tune with the dark matter sceptics – for there are a few. Such was the surprise at Vera Rubin's unexpected discovery in 1978 of the flat rotation curves of galaxies (see previous answer) that a number of people looked for alternative explanations to the idea of dark matter. The most successful was Mordehai Milgrom of the Weizmann Institute in Israel who, in the early 1980s, proposed a theory called 'modified Newtonian dynamics' or MOND.

MOND takes Newton's second law of motion (the one that relates the acceleration experienced by an object to the force applied to it) and says that for extremely small accelerations – such as those experienced by stars orbiting around the centre of a galaxy – the normal law breaks down. That effectively modifies Newton's famous law of gravitation which, in turn, allows galaxies to have flat rotation curves without the need to postulate dark matter. *QED*.

But the first thing any scientist must do with a new hypothesis is test it, and it turns out that MOND is not so easy to test. You might well ask why we can't just investigate MOND's validity by experiment, using weights, pulleys, bits of string and so on – thus validating (or refuting) it. The answer is, however, that the accelerations experienced by objects *anywhere* in the Solar System are too high for the experiment to work. So MOND is kind of defeated by its own delicacy.

There are other reasons why MOND hasn't found its way into the mainstream of scientific thinking as an alternative to dark matter. One is that we have sufficient reason to believe that Newton's laws work as advertised, even at extremely low accelerations. For example, comparing the clumpiness of the original Big Bang fireball (which, as we'll discover in the next chapter, can still be seen) with the clumpiness of today's Universe reveals a 13.7 billion-year process of evolution that is perfectly explained by the laws of gravity as we normally understand them. Moreover, streams of stars in our own Galaxy (as observed, for example in a large-scale star velocity survey currently being undertaken at the Anglo-Australian Observatory) demonstrate motions that clearly support the standard dynamical picture rather than MOND.

Perhaps MOND's biggest problem, though, is that it introduces arbitrary changes into fundamental laws without any good physical basis for doing so. In postulating dark matter astrophysicists are suggesting something self-consistent, as all three types of investigation itemised in the previous answer – together with several others – are telling the same story. So, for the time being at least, it looks as though dark matter is here to stay.

Why do galaxies have spiral arms?

In fact, not all galaxies have spiral arms, most notably the large, rugby-ball-shaped objects we call elliptical galaxies. Some disc-shaped galaxies are also devoid of the elegant spiral structure that everyone thinks of when they imagine these objects.

The origin of the spiral arms is a puzzle that has only been solved within the last few decades. On looking at an image of a spiral galaxy, most of us would draw the obvious conclusion that the arms are strings of stars, trailing along as the galaxy rotates. But a little thought shows that this isn't possible. Given that stars say halfway between the galaxy's centre and its edge will take some 200 million years to go all the way round, how many circuits will they have made in the galaxy's lifetime? The answer is about 60, for a typical 12-billion-year-old galaxy. But stars near the edge of the disc are moving at about the same speed (see 'How do we know there is dark matter?'), so they will take longer to cover the greater distance around their circuit. The end result is that a string of stars would quickly wind up, resembling a clock spring rather than the gently curving spiral arms we're used to seeing.

Another observation rules against the 'strings of stars' idea. If you look at a spiral galaxy in red or infrared light, the spiral arms disappear and you see a rather featureless disc. This is because the spiral arms are made of hot young stars that are blue-white in colour, and blue stars don't show up in red light. The underlying population of stars in the disc are yellow or red stars that are both smaller and older than the blue ones.

This gives us a clue to the true nature of spiral arms. If, instead of a physical string of stars, we are seeing a pattern of vigorous star formation moving through the disc of the galaxy, then the shape of the pattern will be independent of the galaxy's age. In other words, it needn't be tightly wound up. Moreover, the speed of the pattern will be independent of the rotational speed of the stars – it will move through the underlying disc of stars, dust and gas at its own speed, just as the crest of a wave sweeps over bobbing swimmers at the beach.

In arriving at the true mechanism for the origin of spiral arms, astronomers chased a number of red herrings. One theory, popular in the 1970s, was that spiral arms were a product of something called stochastic self-propagating star formation, whose acronym – SSPSF – doesn't exactly roll off the tongue (particularly into a radio microphone). The theory proposed that extravagant young stars quickly squander their hydrogen fuel and blow up as supernovae, triggering a new stellar generation along a 'shock front' of star formation. This shock front is given its characteristic spiral shape by the rotation of the galaxy.

The problem with SSPSF was that it didn't really work, and it was eventually superseded by today's elegant 'density wave' theory. First proposed in the late 1960s by Chia Chiao Lin and Frank Shu, the theory envisages a kind of solitary sound wave moving through the rarefied gas and dust in a galaxy's disc, triggering vigorous star formation as it passes. Once again, the curvature results from the galaxy's rotation. It is thought that the density waves themselves are initiated by gravitational interaction between galaxies.

HUBBLE TROUBLE: STAMPEDING GALAXIES

How do you measure look-back times?

A look-back time is exactly what the name suggests – a measure of how far back in time we're seeing when we look at particular objects in the Universe. It arises – of course – because of the finite speed of light, and it becomes particularly important when we look at very distant galaxies. That's because the light travel time is then a significant fraction of the lifetime of the galaxy. Looking at successively more distant objects – that is, observing them at progressively earlier stages in their development – allows us to investigate the evolutionary processes taking place in galaxies.

But how are look-back times measured? The basic process is surprisingly easy, because the light from galaxies is effectively date-stamped as to when it was emitted. To understand this, we have to go back to the most basic observation we can make about the Universe as a whole – it is expanding. This remarkable fact emerged in the late 1920s, when the American astronomer Edwin Hubble used his own and other observations to deduce that galaxies are flying away from us at speeds proportional to their distances – a relationship now known as Hubble's Law. But rather than thinking 'Was it something we said?', Hubble realised that this means that the Universe as a whole is expanding – and it doesn't necessarily mean we're at the centre. In an expanding Universe, an observer *anywhere* will see galaxies flying away from them.

It was a short step from Hubble's expanding Universe to the deduction that the Universe must have had a beginning when everything was at a single point in space. This was made by a Belgian priest, Georges Lemaître, in an early forerunner of today's Big Bang theory.

As we now understand Hubble's Law, what we're seeing is actually an expansion of space itself, which carries the galaxies along with it. This is what gives us the date stamp on a galaxy's light. Imagine the stars of that galaxy dutifully emitting their light at some time in the past when the Universe was a particular size. By the time their light reaches us, however, the Universe is bigger and the light waves have been gradually stretched – along with the space through which they've been travelling. Longer wavelengths mean the light has become redder. We can measure this so-called 'redshift' with a spectrograph, by analysing the barcode of information contained in each galaxy's rainbow spectrum. The redshift tells us precisely (independently of *any* theoretical models) how big the Universe was when the light was emitted, compared with its size today.

To get from that measurement to a look-back time (and hence, effectively, a distance) depends on a calibration called the extragalactic distance scale – that is, the scale of distances beyond our Galaxy. This involves a succession of different measurements, starting with star parallaxes, and taking in the standard candles provided by pulsating variable stars (see 'How do you measure light years?' and 'Do stars shake?' in Chapter 8). The outcome of this process is a rather well-defined relationship between redshift and distance.

Do galaxies have their own speeds on top of their motion due to the expanding Universe?

If you've read the previous answer, you'll know that the expansion of the Universe is carrying everything along with it. We usually call this the Hubble flow, which is rather a nice way of saying that at any point, space is moving like a river. Now imagine someone in a dinghy on the river. By rowing upstream or downstream they can move with respect to the flow of water around them – indeed, if they like, they can go sideways, too. If there were lots of dinghies in the river, each occupied by a rower doing their own thing, you might say that they all had their own particular motion with respect to the flow carrying them all along.

This is exactly analogous to the situation with galaxies. As well as being transported by the Hubble flow – and thus receding from us – they each have their own particular motion. We usually call them 'peculiar velocities', not because there's anything funny about them, but because they are peculiar to each particular galaxy. Peculiar velocities arise because of the underlying gravity field experienced by each galaxy, and if you can measure them (which is actually rather difficult), you can investigate the forces they are responding to. This allows astronomers to gain further insight into dark matter, because peculiar velocities betray the presence of attracting objects whether they are visible or not.

Do collisions of galaxies happen today?

They certainly do, and at least one is taking place right now in our own Galaxy. It's quite common to find what are euphemistically called 'interacting' galaxies when we look out into the Universe. Perhaps the best known is a pair usually known as 'The Antennae' – although the miscreants' official names are NGC 4038 and 4039. Folks, they're not just interacting – they're colliding in an all-out smash-up worthy of 'Grand Theft Auto'. The reason this pair looks like an insect with antennae is that spiral arms of stars, gas and dust are being thrown off in opposite directions by the impact. Lying relatively close by, at a distance of only 60 million light years, the freeze frame collision of these two objects has been very well studied.

Nearer to home, our own Milky Way Galaxy is in the process of swallowing up some of its neighbours. Being the biggest kid on the block, the Galaxy dominates its immediate environment, drawing in unsuspecting dwarf galaxies and ripping them apart with immense tidal forces. One recent victim is the Sagittarius Dwarf Galaxy, which at present is located on the far side of the galactic centre from the Sun. But its disintegration is revealed by something we call the Sagittarius Stream, which is the trail of stars left behind by the dwarf as it spirals around the centre of our Galaxy in its death throes.

Computer models of galaxy formation demonstrate that this process of swallowing up dwarf galaxies is extremely important in the evolution and growth of large galaxies. The Sagittarius Dwarf is certainly not our Galaxy's first victim, and may even be just the most recent of dozens. One of the problems with such computer models, however, is that they also predict the existence of many more dwarf galaxies than are actually seen around large spirals such as the Milky Way. The deficiency may be due to a faster rate of ingestion than expected or, perhaps more likely, a failure of the many predicted small clumps of dark matter to attract enough hydrogen for the dwarf galaxies to reveal their presence by star formation.

What's a quasar, and are they distributed evenly throughout the Universe?

I usually think of a quasar as a delinquent young galaxy although, in reality, the galaxy is just the outlying structure that plays host to a recklessly energetic object at its core – the quasar itself. Fortunately, quasars seem to be extinct today, as we only see them at great distances. Even the closest has a look-back time of more than a billion years.

When they were discovered in the 1960s, quasars were a mystery because they seemed to emit far more energy than was possible for such extremely compact objects. The name derives from 'quasi-stellar object' (or, more accurately, 'quasi-stellar radio source', since the first ones to be found revealed themselves by their strong emission of radio waves). So, they looked like stars – but were far too distant to be stars. And rather than confining themselves to one particular waveband (such as visible light) their energy output was spread over a broad range, from X-rays to radio waves.

Moreover, this radiation was found to vary on timescales of hours or days, which set limits to the size of the energy-emitting region. Anything bigger than a few light hours or light days would blur out the variation, because of the radiation travel time from front to back. (A light hour, by the way, is about a billion kilometres, and a light day is 24 times further.)

We now believe that the quasar phenomenon is at the more frantic end of a broad spectrum of activity in the cores of galaxies, and is produced by material being swallowed up by a super-massive black hole far larger than the one at the centre of our Galaxy (see 'Where is the nearest black hole?' in Chapter 8). Black holes with masses up to 100 million times that of the Sun may be responsible for quasars, as they accumulate around themselves a rapidly rotating 'accretion disc' of material, which they eventually consume. It is turbulence within the accretion disc that causes the prolific energy output.

Quasar activity must be associated with youth in galaxies, because we only see them in the distant Universe – that is, in earlier eras. It is unclear what triggers the phenomenon, and it may be that for any given galaxy, quasar activity lasts for a relatively short time. Like galaxies, quasars concentrate together, revealing a characteristic clumpiness in the way matter is distributed on very large scales. But to astronomers, quasars have one singular advantage over galaxies. They are without doubt the most luminous objects in the Universe, shining brightly across vast tracts of space from the very distant past. They provide a vital and effective means of probing the intervening space in great detail.

CHAPTER 10

INDUSTRIAL-STRENGTH ASTRONOMY

COSMOLOGY AND BASIC PHYSICS

When you come to think of it, astronomy is a pretty daft business. Here we all are, insignificant microbes on the surface of a cinder left over from the formation of the Sun, trying to understand a Universe whose age and scale simply boggle our insignificant microbe brains. No wonder astronomers are often considered eccentric. What is even more absurd is the idea that we think we understand how it all works. A Big Bang, some 13.7 billion years ago, to start it all off, then everything else following with almost pre-ordained inevitability – from galaxy formation and supernovae to planet Earth and *The Simpsons*. This is big-time audacity on the part of the microbes.

We've even given this audacity a name – cosmology. It's defined as 'the study of the origin and evolution of the Universe as a whole'. And, for a bunch of microbes, we do seem to be doing rather well at it – despite the fact that it's definitely at the industrial-strength end of astronomy in terms of its demands on our understanding.

Why do we have such confidence in our ideas on the origin and evolution of the Universe? There are several reasons. On the observational side, as astronomical technology has improved, scientists have been blessed with copious quantities of new data. Bigger telescopes, better instrumentation, more advanced spacecraft – together with a greater understanding of fundamental physics, they have strengthened our cosmological models. On the theoretical side, these models have become more sophisticated as a result of the efforts of the super-brains – the Einsteins of our own generation. Hawking, Rees, Peebles, Sandage, Silk, Longair – these names are legendary to today's workers in the field. And overall, a remarkable self-consistency has emerged in our understanding of the history of the Universe.

As we saw in the last chapter, information on the Universe's origin and evolution can be drawn from a number of sources, including the history of our own Galaxy and other, nearby galaxies. Such studies are sometimes known as 'near-field cosmology', since these objects can be studied in great detail, and accurate inferences made about the processes that have taken place during their lifetimes. They act as models for more distant objects, whose evolution can then be better understood.

It is studies in the far field that most of us think of when we talk about cosmology, however, and that means investigations into the most distant objects. These include very remote galaxies, highly energetic quasars and events called gamma-ray bursts, which are thought to be giant stars exploding at the end of their brief lives. Because of their vast distances and the finite speed of light, we see such objects as they were billions of years ago, allowing us to look directly into the past. Far-field cosmology also takes in gravitational lensing (a handy tool for exploring the distant Universe, described in the last chapter) and studies of the oldest and most distant thing we can observe – the cosmic microwave background radiation, or CMBR.

This CMBR became the clinching evidence for the Big Bang model of the Universe when it was accidentally discovered in 1965 – by two engineers at the Bell Telephone Laboratories. They had found a weak signal in the microwave region of the radio spectrum that seemed to be coming from everywhere in the sky. It turned out that the whole sky was glowing as if it had a temperature of 2.7 degrees above absolute zero – and that something like this had been predicted almost two decades earlier by the physicist George Gamow. Today, the CMBR is beginning to yield up its secrets, and has huge potential for increasing our understanding of the Universe. So what is it, and how did it get there? And why is it so important?

To answer these questions, we have to start with the Big Bang theory itself. The name, incidentally, came from a rather derogatory remark made by the late Sir Fred Hoyle (a proponent of the rival Steady State theory) in a lecture series on BBC radio in the late 1940s, later published as *The Nature of the Universe*. But the theory originated in work by Alexander Friedmann and Georges Lemaître back in the 1920s. Friedmann solved the mathematical equations of Einstein's general relativity to reveal the possibility of an expanding Universe (see 'Does gravity bend space?'), while Lemaître realised that expansion implies an initial creation event. In other words, at some time in the very distant past, the whole of space was compressed into a single point.

Today's theoretical picture of the Big Bang is a vastly more sophisticated affair. The so-called 'standard model' includes the following basic ingredients:

1. A hot Big Bang explosion 13.7 billion years ago, in which not only matter, but also space and time were created.
2. An extremely brief period of violent inflation, during which the infant Universe expanded by at least 10^{30} times while it was about 10^{-35} of a second old; followed by the steady expansion still taking place today.
3. The existence of copious quantities of hidden material in the form of unknown particles (the dark matter described in the last chapter).
4. A gradual slowing-down of the expansion by the mutual gravitational attraction of all the matter in the Universe.
5. A cosmic repulsion, or vacuum energy (usually called dark energy) that today seems to be causing the expansion to *accelerate*.

The result is an intoxicating brew that intrigues cosmologists with its various uncertainties, spurring them on to devise clever new observations to firm up their ideas.

Cosmologists themselves cite three basic observations as the model's main pillars of support. First, the Universe is indeed expanding, as first demonstrated by Edwin Hubble in 1929. Then there is the fact that light elements such as helium and deuterium are found in the ratios predicted by the Big Bang model. Finally, and perhaps most dramatic, there is the existence of this cosmic microwave background radiation.

To understand the CMBR, you have to imagine the Universe when it was only a few hundred thousand years old. While some of the weirdest episodes of its early history – inflation, for example – were long past, it was a very different place from today's Universe. Space itself was still glowing brilliantly like an opaque fog of radiation – a fireball. There was no means of seeing through it.

Then, quite suddenly, when the Universe was about 380,000 years old, it became transparent. There's no need to go into the details of why this happened, but it took place everywhere over a relatively short period of time. Of course, the Universe is still transparent, and because of this – and because of the finite speed of light – we can now look so far back in time that we can see the instant when the fog cleared. And what do we see? Yes, the ubiquitous cosmic microwave background radiation.

There is a well-known and rather neat analogy to explain why we can still see this cosmic fossil. Imagine yourself at an outdoor concert in the middle of a huge field crammed with people (I'll leave you to decide what kind of music you're listening to...). At the end of the performance there is wild and enthusiastic applause, but the band (or whatever) eventually gets fed up with the adulation, and signals an end to the cheering. Instantly, everyone in the audience falls silent. Now, do you immediately hear silence?

The answer is, 'No', because even though you and everyone else around you have stopped cheering, sound from your surroundings is still on its way to you. One second after the applause stops you'll still hear the sound made by people 330 metres away just as they stopped cheering; a second later you'll still hear sound that came from people 660 metres away – and so on. You are in the centre of an expanding circle of silence – a circle expanding at the speed of sound.

Similarly, we are at the centre of an expanding sphere of transparency in the Universe, and we are still receiving radiation that was emitted from the glowing fog almost 13.7 billion years ago. This expanding sphere is sometimes called the 'last scattering surface'. I suppose its equivalent in the terrestrial analogy might be called the 'last cheering circle'.

There is one important ingredient missing from the cheering analogy. In that instance, the sound you hear is at the same pitch as it was emitted. But this is not the case with the CMBR. When the radiation was emitted, it was brilliant white light, the product of a fireball with a temperature of several thousand degrees. Today, we perceive it as weak radiation in the microwave region of the spectrum – short-wavelength radio waves. This is because the Universe is expanding, and as it moved through space the background radiation was stretched in wavelength – by about 1300 times. It's the equivalent of more than a thousand-fold drop in the pitch of the sound waves.

The outdoor concert analogy can explain one other attribute of the CMBR. Suppose the audience includes clumps of people who are slightly more or slightly less enthusiastic about the band than the average. When the cheering stops, some directions will have slightly louder applause coming from them than others. These directions might well vary as the area of silence expands, but the basic idea remains – there will be structure in the 'last cheering circle'. This is exactly what happens with the CMBR. Even though it is exceedingly uniform over the sky, it does have faint ripples – in fact, at the level of about one part in 100,000 – and it's just as well it does. Otherwise, folks, we wouldn't be here.

The ripples in the CMBR take the form of slight variations in the temperature of the fossil radiation. They arise from variations in density, gravity and velocity in the fireball all those billions of years ago, and are the result of sound waves propagating through it – the 'bang' of the Big Bang, if you like. It was the British-turned-American astronomer Geoffrey Burbidge who pointed out in 1980 (before the ripples were detected) that they were the gravitational seeds of the earliest galaxies – and if they weren't there, neither were we...

The story of the ripples' detection by the *COBE* spacecraft (COsmic Background Explorer) in the early 1990s is well known. The CMBR anisotropy (its departure from uniformity) was measured by careful analysis of the *COBE* data by a team that included Charley Lineweaver – now an astronomer at the Australian National University – and revealed a treasure-trove of cosmological information locked up in the ripples. Much more detail has been revealed in the last few years by another spacecraft called *WMAP* – the Wilkinson Microwave Anisotropy Probe – which was launched in June 2001. This spacecraft has measured the temperature of the cosmic background glow to about one part in a million (that is, a few millionths of a degree) on angular scales down to 0.3 degrees over the whole sky. Soon, there will be another, even more powerful spacecraft investigating the CMBR – ESA's *Planck* mission, due for launch at the end of 2008.

It's the analysis of the characteristic scale of these hot and cold spots (in degrees on the sky) that has revealed priceless cosmological data, including the age of the Universe, the time at which the glowing fog cleared, the reality of inflation and the effect of dark energy, as well as details of conditions within the fireball.

As we discovered in the last chapter, the level of clumpiness found in today's Universe – the tendency of galaxies to group together in clusters or strings – is entirely consistent with the measurements of the ripples in the CMBR. In other words, the 'seeds of creation' scattered through the Universe some 380,000 years after the Big Bang are still there today, frozen into the pattern of galaxies we see around us. Such self-consistency within observations made by a wide range of different techniques is very encouraging to cosmologists.

There are, however, still a few doubts in our interpretation of the evidence for the standard Big Bang model. One is the recent discovery of the provocatively named 'axis of evil', an apparently preferential direction in the CMBR's ripple pattern – even though theory says it should be completely without structure on large scales. This has led some cosmologists to question whether we are missing something. As one British astrophysicist, Peter Coles, recently put it, 'when generating theoretical ideas scientists should be fearlessly radical, but when it comes to interpreting evidence we should all be deeply conservative' (*New Scientist*, No. 2593, 2007).

Despite such misgivings, there's no doubt that cosmology has progressed enormously since, only a few decades ago, it was regarded as a branch of philosophy. Today's industrial-strength version really does have a lot to recommend it. All in all, it's not a bad effort for a bunch of microbes.

LET'S GET PHYSICAL: TOOLS OF THE COSMOLOGIST

Can anything travel faster than the speed of light?

The simple answer to this question is, 'No', but there is an underlying subtlety that once caught me out on an early-morning phone-in show, when my brain wasn't properly on-line. Let's deal with the straightforward bit first. It is perfectly true that nothing can move faster than the speed at which light travels in a vacuum – that is, through empty space. This speed is usually denoted by the symbol c and, as we saw in Chapter 8, it has a value of about 300,000 km/second. Incredibly, breathtakingly, preposterously fast – the ultimate speed limit.

In physics, c assumes an almost mystical significance because no matter how fast the source of light is moving, light *always* travels at this speed in a vacuum. This flies in the face of common sense, but it was hinted at in the work of James Clerk Maxwell during the 1870s, and demonstrated experimentally in 1887 by the physicists Albert Michelson and Edward Morley. The invariance of the speed of light then found its way into the Special Theory of Relativity, Albert Einstein's theory of motion developed in 1905. Special relativity, in turn, proved that in order to accelerate any object to the speed of light, you need to supply it with infinite energy – which is impossible. Thus you can never travel at, or faster than, the speed of light. This, of course, is a great shame, given the distances faced by any would-be explorer of the Universe.

By supplying enough energy to something very small – such as a subatomic particle – it can be made to travel at almost the speed of light, and this is what happens in the particle accelerators beloved of nuclear physicists. Cosmic rays – naturally occurring particles that constantly bombard us from space – also travel at speeds close to that of light. And it is here that the subtlety arises.

My early-morning questioner asked whether there was any equivalent in light of the sonic boom generated when an aircraft exceeds the speed of sound in air. In response, I outlined the above, emphasising that nothing can travel faster than light – and then moved on to the next question. It was left to another listener to point out politely by email that while this is true in a vacuum, there are other circumstances in which it is not. And, indeed, there is an exact light-wave analogue of the sonic boom. Oh, the shame of it... so the following week I provided an explanation, giving credit to listeners more wide-awake than me for keeping me honest.

The fact is that as a cosmic ray particle moving close to the speed of light (or 'at a relativistic velocity', to use the trade term) approaches the Earth's surface, it encounters air. So it's no longer moving through a vacuum. The importance of this is that the speed of light in any transparent medium is *lower* than that in a vacuum. Light travelling through your bathwater, for example, is moving at only three-quarters of its vacuum speed, some 225,000 km/second. (The bathwater vacated by my sons when they were youngsters sometimes didn't permit the passage of light at any speed, but that's another story.) In air, the difference is much less, but the speed of light is still appreciably slower than that in a vacuum.

So here's the logic. A cosmic ray particle travelling at a relativistic velocity encounters the Earth's atmosphere and suddenly finds itself moving faster than the local speed of light. And, exactly like a supersonic jet, it emits a shock wave. Not in the form of a sonic boom, of course, but as a cone of radiation in the visible region of the spectrum. A beam of light, in other words. This faint light is known as Čerenkov radiation, after the Russian physicist who discovered it in 1934, and it was of pioneering importance in early cosmic-ray studies. It is also the mechanism that causes some objects in deep space – certain types of gas cloud, for example – to emit light. It is *not*, however, something to be trifled with when your brain is half-asleep first thing in the morning...

What is spacetime?

For over two hundred years, we were told by Isaac Newton that we occupied a Universe of three dimensions, and then along came Einstein who said no, there's actually another one. Einstein was not the first to suggest that time might behave like a fourth dimension, but the idea is so deeply ingrained in his Special Theory of Relativity that we tend to associate it with him. It was, in fact, Hermann Minkowski who gave us the full-blown four-dimensional Universe in elegant mathematical form in 1908.

The difference between Newton's Universe and Einstein's is that Newton thought of time as an absolute – something that was regularly meted out, and against which everything else was measured. There were no difficult questions about simultaneity, for example – if two events happened at the same time anywhere in the Universe, they were simultaneous. But

Einstein asked what would happen when simultaneous events were seen by two observers, one of whom was moving with respect to the other. It turns out that because of the finite speed of light, the notion of simultaneity gets rather vague.

It's easier to visualise this paradox if you swap the notion of time for that of space – because another feature of this brave new world is that the four dimensions are subtly interchangeable. So there you are, sitting on the upper deck of your brand-new Qantas Airbus A-380, *en route* from Sydney to London. You've finished reading this morning's crumpled *Sydney Morning Herald* and given up on the Sudoku, and dinner is about to be served. Half an hour later, as you take your last sip of lukewarm coffee, you reflect on the fact that two events – the beginning and end of your meal – have occurred in exactly the same place, since you haven't moved from your seat. But to an observer on the ground, they are separated by nearly 500 kilometres. The occurrence of events in space and time is therefore relative rather than absolute – hence the name of Einstein's theory. And it's 'special' because it refers to the special case of things moving with respect to one another at constant speed, as in an airliner at its cruising altitude.

This notion of spacetime is one of three basic postulates of special relativity. The other two are that the laws of physics are the same whether you're moving or not (airline coffee stays in the cup – assuming there's no turbulence – even though you're moving at nearly 1,000 km/h over the ground); and that the speed of light in a vacuum is always the same, no matter how fast the source of light is moving. As we saw in the previous answer, that surprising fact was first demonstrated back in the 1880s.

When Einstein worked his mathematical magic on these ideas, he got some unusual results about time and space that take on real significance for things moving at speeds close to the speed of light (see 'Is it true that clocks slow down when you travel very fast?'). But more dramatically – and entirely unexpectedly – he arrived at perhaps the most famous equation of all time, linking the mass of an object with its value in energy. You know the one I mean...

What does $E = mc^2$ mean?

Oh, all right. I wasn't actually going to quote it, but it seems to have muscled in. No doubt it's something to do with being the most famous equation in the world. As we saw a couple of answers back, c is the speed of light. It's a very large number. If you multiply it by itself (i.e., c^2), then it becomes an *incredibly* large number. The m in the equation is the mass of an object, and E is what you'd get if you could convert that object entirely into energy.

So, because its mass is multiplied by this enormous c^2 number, the energy value of any object is absolutely huge. There are certain processes that allow you to make such a total conversion of mass into energy, most of which take place when you start tinkering with atoms at the level of their nuclei. It is this equation, therefore, that allows us to understand the prodigious energy output of stars, and to predict the devastation associated with nuclear weapons. A bit of a two-edged sword, really. Fundamentally, the equation tells us that mass and energy are truly equivalent, which has applications over a wide area of physics.

Does light have mass?

Back in Chapter 2 ('Why do you build telescopes on mountain tops rather than in flat deserts?'), we saw that light can be described as a wave motion, but modern physics regards it equally well as a stream of particles called photons. A photon's energy is related inversely to its wavelength, in the sense that higher-energy photons correspond to light of a shorter wavelength. So – because mass and energy are equivalent (see the previous question), yes, light does have mass. Where it differs from ordinary particles is that it doesn't have a rest mass – that is, its mass when at rest with respect to the observer.

Is it true that clocks slow down when you travel very fast?

Bizarre though it sounds, this is indeed true. Once again, it's a consequence of Einstein's Special Theory of Relativity. There are some subtleties to the effect – in particular, that it only becomes appreciable when you're travelling close to the speed of light. The effect is called time dilation, and it refers specifically to the way clocks behave when viewed by observers in different states of motion. One of the observers is usually considered to be stationary and, for some reason, I always think of a man sitting in an armchair with a cup of tea. Don't ask me why. I have noticed, however, that he looks a bit like Einstein.

Let's imagine that the stationary observer has a clock beside him, but that he can also see another identical clock that's going past him on a very fast train. What he will see between sips of tea is that the clock on the train is ticking more slowly than the one beside him. In other words, its time is dilated. Paradoxically – because the main point about relativity is that all motion is relative – a woman beside the clock on the train will see the armchair man's clock going slower than hers. The point here was rather neatly expressed by Stephen Hawking, who wrote that relativity 'required abandoning the idea that there is a universal quantity called time that all clocks would measure. Instead, everyone would have his or her own personal time'.

The time dilation effect is frequently observed in physics, and has been checked many times. For example, back in 1940, the lifetime of a subatomic particle called a muon was measured in the laboratory to be about two millionths of a second. However, muons were known to last very much longer than this when they arrived in the Earth's atmosphere as cosmic rays (see 'Can anything travel faster than the speed of light?'). That difference was deduced correctly to be due to the muons' decay time being dilated by their very high velocities.

By the way, special relativity also tells us that a stationary observer will see a moving object getting slightly shorter in the direction in which it's travelling. Once again, the effect is only significant at speeds close to the speed of light. If you want to look thin, it's one way of doing it, I suppose.

Does gravity bend space?

To be exact, gravity *is* bent space. But to understand why, we have to look at the best theory of gravity available and that, once again, is due to Albert Einstein. It began in 1907, when the 28-year-old Einstein had what he later described as '*der glücklichste Gedanke meines Lebens*' – 'the happiest thought of my life'. Wow – Einstein's happiest thought. What on Earth could that be about?

In fact, the idea was off the planet rather than on it. Einstein was thinking about how his special relativity theory might modify Newton's famous theory of gravitation. Arguably the greatest intellectual feat of all time, Newton's theory had proved dazzlingly successful in explaining the motions of objects in the Solar System right down to the finest detail. Just one minuscule aspect of Mercury's orbit didn't seem to fit, but everything else did. So, mused Einstein, how would relativity alter it, given that Newton's gravity took no account whatever of time, and Einstein now knew that time and space were intimately linked?

For some bizarre reason known only to Einstein, he then imagined himself falling from the roof of a house, and realised that during this rather inconvenient event, he would feel no gravitational force. He would certainly feel something when he hit the ground, but that doesn't matter in a thought experiment such as this. Einstein reasoned that he would be in a state of freefall, and if his pipe fell out of his mouth or coins fell out of his pocket, they would appear to float around him as if there were no gravity. We fortunate occupants of the twenty-first century are used to the idea of astronauts experiencing weightlessness, but in Einstein's day it was an entirely novel concept.

This quickly led Einstein to the next step – the realisation that the effects of a gravitational field and an applied acceleration are identical. When you're falling off a roof, one cancels out the other and you become weightless – for a while. If you were sitting in a windowless compartment on a rocket in space, and someone lit the fuse, you would be unable to know whether the force you felt was due to the rocket's acceleration, or to gravity. Therefore – locally at least – they must be the same thing.

This happy thought of Einstein's was a major breakthrough, now called the Principle of Equivalence. In fact, it wasn't until 1912 that he set out a formal statement of the principle, putting it in terms of the equivalence of an object's gravitational mass (the way it responds to the pull of gravity) and its inertial mass (the way it responds to a force such as the thrust of a rocket). Einstein then soared in his thinking, bringing together the mathematical tools needed to develop a new theory of gravity based upon the idea of accelerated reference frames. (Imagine yourself back in the Airbus in 'What is spacetime?' When it makes its take-off run with those four giant turbofans at full power, you and everyone else on board are in an accelerated reference frame. It feels great.)

Einstein quickly realised that much of the mathematical development of the theory boiled down to a problem of geometry – because the laws of relativity that apply in ordinary, or so-called Euclidean, space don't work any more. But if you invoke a geometrical construction called a four-dimensional Riemannian manifold, then the accelerated reference frames become as docile as kittens. Well, nearly – the mathematical formulation is still exceedingly difficult. You might be forgiven for thinking, as I once did, that a Riemannian manifold is something you'd find under the bonnet of your car, but it's merely a particularly complex model of space developed in the 1850s by a gifted German mathematician called Bernhard Riemann.

What followed was Einstein's General Theory of Relativity, a startling new theory of gravity, which made the wildly improbable assertion that spacetime itself can bend, warped by the presence of matter. Matter, in its turn, responds to the distorted geometry of spacetime by moving within it. The theory also predicted that time is slowed when you accelerate – or, equivalently, when you are in a gravitational field. This so-called gravitational time dilation is analogous to the time dilation effects of special relativity (see the previous question).

The new theory was stunning in its implications. Shortly before submitting his description of it for publication, in November 1915, Einstein realised that it exactly accounted for the observed anomaly in Mercury's orbit. 'For a few days, I was beside myself with joyous excitement', he wrote later. And the dramatic verification of his theory's prediction that the Sun's gravity bends light rays, using observations of the total solar eclipse of 25 May 1919, is now the stuff of legend. As *The Times* of 7 November 1919 blared: 'Revolution in science – New Theory of the Universe – Newtonian ideas overthrown'. None of which was any exaggeration.

More than 90 years after its formulation, general relativity is still the best theory of gravity we have. It has survived all the tests that have been thrown at it and has successfully predicted the existence of black holes, the expanding Universe, gravitational lenses, gravitational waves, and so on. It underpins cosmology. Most of us are familiar with the idea that spacetime is curved, and Einstein is regarded as the greatest genius of the twentieth century, not to mention a few other centuries besides.

The one thing that general relativity can't cope with is gravity on the very smallest scale, since the theory is based on spacetime as a continuous medium rather than a succession of distinct steps or 'quanta' – and we know that this is how the sub-microscopic world works. The search for a theory of quantum gravity is well under way... but that's another story, folks.

Does gravity have a speed?

Once again, yes, it does. Moreover – and quite remarkably – a recent experiment succeeded in measuring its speed. As we saw in the previous answer, space (or more correctly, spacetime) bends in the presence of matter to produce gravity. But it would be wrong to imagine space as being flexible like a bendy toy, or a pipe-cleaner. In fact, it is incredibly rigid – billions of times more rigid than steel. Nevertheless, general relativity predicts that waves will travel through the fabric of space, rather like ripples on a very rigid pond.

These waves have their origins in gravitational disturbances such as supernova explosions and colliding stars, and the theory predicts that they will travel at the speed of light. Thus, if the Sun was suddenly removed from the centre of the Solar System, the Earth's orbit would be unaffected until the gravitational wave from the Sun's demise arrived eight minutes later – whereupon a rather surprised Earth would head off in a straight line.

Space is so rigid that gravitational waves are exceedingly difficult to detect, and technology is still a few years away from picking them up. However, we know with certainty that they exist, thanks to the 1974 discovery of a so-called binary pulsar, whose energy loss can only be explained by the emission of gravitational waves. That finding led to the 1993 Nobel Prize for Physics for its discoverers, Joe Taylor and Russell Hulse. Once gravitational waves can be routinely detected and measured, we will have an exceptionally powerful new technique for probing the Universe. It may even allow us to study the mechanisms of the Big Bang itself.

The speed of gravitational waves was measured back in 2002, when the planet Jupiter passed in front of a distant quasar. Like any other massive object, Jupiter bends the trajectory of light passing close to it, and two American scientists, Ed Fomalont and Sergei Kopeikin, used this bending to estimate the speed of the waves. They were able to do that because Kopeikin (a theoretical physicist) found he could reformulate the equations of general relativity in such a way that gravity became analogous to electromagnetic radiation. The answer they got – within the accuracy limits of their experiment – was that, yes, Einstein was right. Gravity does travel at the speed of light.

ONE HELL OF A BANG – THE STANDARD MODEL AND BEYOND

What was wrong with the Steady State theory?

Some readers of this book may remember the intense scientific controversy that raged in the 1950s and 1960s between the Big Bang and Steady State theories of the Universe. The Steady State theory was proposed by Thomas Gold, Fred Hoyle, Hermann Bondi, Jayant Narlikar and others, and said that while the Universe is indeed expanding, it had no beginning and has had the same properties at all times. Thus, matter is continuously being created to give a constant average density. The theory avoided the problems associated with a Universe expanding from a single point (see the next answer, for example). It also got rid of an embarrassing situation in which the Universe appeared to be younger than some of its contents – as was suggested by early estimates of its age from the Big Bang theory.

The two observations that toppled the Steady State theory were the 1965 discovery of the cosmic microwave background radiation, and the subsequent finding that galaxies and quasars look different as you observe progressively more distant samples – suggesting strongly that they evolve with time. For the reasons outlined in the introduction to this chapter, the Big Bang theory is almost universally accepted today. However, the Steady State theory was of great value in stimulating cosmological investigation, and also provided a framework for some of the more subtle aspects of the Big Bang theory.

What was there before the Big Bang?

The smarty-pants answer to this is that the last thing heard before the Big Bang was someone saying 'Oh, bugger!' The real answer is even harder to get your head around. In the standard model of an expanding Universe, time starts at the instant of the Big Bang. Thus, the concept of 'before the Bang' is meaningless – time didn't exist. I think it was Stephen Hawking who drew the analogy of travelling North on the Earth's surface. It's fine to talk about going northwards – until you get to the North Pole. Then where do you go? Northwards is no longer defined, and neither is time before the Big Bang.

Some eminent physicists – such as Gabriele Veneziano of the CERN laboratory in Geneva – believe that the idea of the beginning of time is a myth. They point to the weakness in general relativity (on which the standard model is based) mentioned above in 'Does gravity bend space?' This is that relativity doesn't work on very small scales, when so-called quantum effects come into play. At that level, things move in small jumps rather than smoothly, and relativity has no answers. Clearly, if the Universe was very small at some instant in the very distant past, quantum effects were vitally important. One of the most fashionable of today's quantum theories is String theory, which says that at the most basic level, subatomic particles aren't actually particles, but vibrating strings. String theory predicts that time didn't have a beginning, so even though the Big Bang still took place, the Universe may have predated it.

What's outside the Universe?

Once again, the standard model provides the answer to this. Nothing. Not even empty space. All of space is, by definition, within the Universe, so there can't be an outside. What is perhaps more alien to our thinking is the idea that the Universe has no edge – no boundary where space stops. That's because we intuitively think of space on very large scales as being similar to the way we perceive it in our local environment – well-behaved, and with parallel lines never meeting. But, in fact, on those large scales, space is curved – and parallel lines do meet.

The easiest way to visualise this is to think of the surface of a sphere as being a two-dimensional representation of the Universe. Even though its surface area is finite, it has no boundary. You could keep going for ever and not come to an edge. In fact, space is probably curved in a rather different way, but the general principle remains the same.

Rather than imagine an edge to the Universe, astronomers often think in terms of horizons beyond which we can't see. The cosmic microwave background radiation described in the introduction to this chapter is the most obvious of these. Because of the extremely rapid inflation of space immediately after the Big Bang, however, the Universe itself is very much bigger than the volume of space defined by that horizon, and may even be infinite.

Where was the Big Bang?

It sounds a bit glib, but the answer to this question – once again courtesy of the standard model – is 'everywhere'. When we think of something exploding, we nearly always imagine it exploding into space, because that's what we're used to seeing. Usually on our TV news bulletins, from one of the world's trouble spots. But in the Universe, all of space has expanded from a single point – in fact, a singularity rather like a black hole (see 'Where is the nearest black hole?' in Chapter 8). This means that at the beginning, no one part of the Universe was any nearer to the centre than any other part. Thus the answer, 'everywhere'.

A related issue that has occasionally been raised by listeners is, 'If the Universe is expanding, where does the extra space come from?' The standard model doesn't really try to explain what space is, and just says it gets bigger. Some quantum theories, however, envisage space as a kind of foam of quantum particles – whatever they are – so clearly there needs to be a mechanism for this kind of space to expand. The issue is also bound up with the hypothetical causes of dark energy, discussed later in this chapter.

If the Universe is 13.7 billion years old, how can anything be further away than 13.7 billion light years?

There are two aspects to this question. First, a reminder that everything in the Universe observable with current techniques is within a look-back time of 13.7 billion years, since at that time in the past, space was glowing brightly. We can still see that limit as the cosmic microwave background radiation (CMBR). You'll notice that I've spoken of a *look-back time* of 13.7 billion years rather than a *distance* of 13.7 billion light years. That's because the true distance to the horizon of the CMBR is much greater than 13.7 billion light years – since the Universe has been expanding through the intervening time.

In fact, the CMBR is probably more like 35 billion light years away in what is called 'proper distance' – that is, the distance it would be if we could see the whole of the Universe as it is today. Of course, we can never see that, because all information comes to us at the speed of light, so it's much more common to think of the distance to the CMBR as 13.7 billion light years – which is technically known as its 'comoving distance'.

Even when that is explained, however, there is still a puzzle. In the last question but one, I mentioned that the Universe itself is very much bigger than just the bubble defined by the distance to the CMBR (which, by the way, is always centred on the observer, no matter where they are in the Universe). Does that mean that the Universe has expanded faster than the speed of light? The answer is, yes – during the brief inflationary period that immediately followed the Big Bang (see the introduction to this chapter).

The reason why this doesn't violate the speed limit defined by Einstein's special relativity (see 'Can anything travel faster than the speed of light?') is that it's not the speed of something moving *through* space, but the expansion speed of space itself. Space is the reference frame, and it can expand at any speed it likes. Things moving through it, however, can't exceed the speed of light.

What is dark energy?

This is of one of the most exciting questions facing cosmologists today. As recently as the mid-1990s, we believed that the expansion of the Universe was gradually slowing down as a result of the mutual gravitational attraction of everything within it. An entirely reasonable assumption, you'd think, but a few heretical souls in the astronomical community had some doubts about this model. They wondered whether a kind of dark energy, or 'springiness' of space, might be starting to counter the deceleration, pushing the Universe into an era of accelerated expansion. It was a rather esoteric point until, in 1998, two separate groups of scientists produced hard evidence that the Universe is, indeed, expanding more rapidly today than it was seven or eight billion years ago.

This evidence came in the form of observations of a particular kind of supernova – the so-called Type Ia – at very great distances. These supernovae are caused by old stars exploding violently as a result of matter from a nearby companion star being deposited onto them. They provide extremely bright standard candles, easily outshining their host galaxies for a few days. What caused all the excitement was that these remote supernovae were dimmer than they should have been, given their estimated distances – and hence look-back times – from our Galaxy.

This suggested that, yes, the expansion of the Universe was accelerating, a result that has now been confirmed by a number of different methods. Moreover, we now know that the dark energy driving the acceleration is the largest single component of the Universe, amounting to some 75 per cent of its total mass/energy density. So what is it? When consideration was given to this question, an intriguing possibility emerged.

Some fourteen months after completing his General Theory of Relativity, Einstein had realised that his equations represented a Universe that must either expand or contract – it couldn't remain static. This was long before Hubble's 1929 discovery of the real Universe's overall expansion, so Einstein thought he had a serious problem on his hands. To fix it, he introduced a new element into his equations, which he called the 'cosmological constant' (conventionally symbolised by the Greek letter capital lambda, Λ).

This new quantity would represent a property of space itself, allowing an attractive or repulsive force to be introduced at a constant rate throughout the whole Universe, compensating its natural tendency to expand or contract. The value of Λ would have to be determined by observations, of course. The problem duly sorted – mathematically, at least – Einstein sat back feeling rather pleased with himself. However, as soon as the expansion of the real Universe was discovered, he quickly withdrew the idea. Years later, the great Ukrainian-American physicist George Gamow disclosed that 'when I was discussing cosmological problems with Einstein, he remarked that the introduction of the cosmological term was the biggest blunder he ever made in his life'.

Once Hubble and others had established the expansion of the Universe, most cosmologists simply assumed that the value of Λ was zero, and forgot about it. Then, in the late 1990s, when astronomy was overwhelmed by the euphoria of the newly found accelerating Universe, the obvious question was asked. Could the dark energy be something to do with that long-neglected orphan of general relativity, the cosmological constant? If it had the form of a negative pressure, and Λ had a value small enough that it only began to overcome gravity when the typical distances separating galaxies had become very great, then it might just fit the bill.

There are other theoretical possibilities, but they require the introduction of various flavours of 'new physics' – phenomena such as quantum gravity and String theory that are not predicted by relativity. They can be summarised as:

1. A new fundamental force, the whimsically named 'quintessence', which echoes the fifth element of ancient Greek philosophy. Like the cosmological

constant, this would have to be a dark energy with negative pressure, but a key difference is that quintessence evolves with time, leaving its imprint on the Universe only in the relatively recent past.

2. A modification of gravity, which makes it act differently on small scales from the way it acts on large scales. If correct, this could eliminate the need for an underlying negative pressure.
3. Abandoning the cosmological principle, which postulates that the Universe is the same in all directions. This would permit a Universe that has significant differences between one place and another, perhaps again eliminating the need for an overall repulsive pressure.

No one really knows what causes dark energy. The best guesses involve a seething foam of virtual particles at the quantum level, popping in and out of existence and imbuing the fabric of space with negative pressure. Theories on this bizarre area of quantum physics are many and various – and far from complete. Embarrassingly, even the best of them predict a repulsive force that is 120 orders of magnitude (yes, 10^{120}) bigger than what we observe. Such intense repulsion would tend to rip everything apart, and certainly would have prevented atoms from forming in the early Universe. This estimate is famously considered by many to be the worst prediction in the whole of science.

A good start on trying to understand the problem, however, might be to identify which model of dark energy best fits the astronomical observations. Cosmological constant or quintessence – or something else? The problem is that most of the required observations don't yet exist. A number of groups throughout the world are now actively engaged in the process of obtaining them, typically by extending supernova observations to greater distances and greater numbers of objects. But the best chance of really tying down the intimate details of dark energy comes from more subtle – and more difficult – observations.

The nature of dark energy has a significant influence on the large-scale geometry of the Universe, just as dark matter does. Therefore, if this geometry can be probed at different stages in the Universe's history, there is a real chance that the correct model of dark energy can be identified, and its characteristics determined precisely.

We already have an accurate knowledge of the large-scale geometry at two key times in cosmic history. One is in the very early Universe. Detailed observations of the CMBR have revealed what the Universe was like in its infancy. And large-scale surveys of the three-dimensional distribution of galaxies in *today's* Universe have been made with the 3.9-metre Anglo-Australian Telescope and other facilities. The missing ingredient is a similar three-dimensional survey of galaxies at great enough distances that they correspond to a look-back time of about half the age of the Universe – seven billion or so years ago – when dark energy first began to make its presence felt.

To do the job properly requires a survey of around a million faint galaxies, a hugely ambitious programme. This is currently in the planning stage, and the details are yet to be finalised. It's not even clear which of the world's great telescopes will carry it out. But the new dark energy survey – and its exciting outcomes – are something we can all look forward to during the second decade of the twenty-first century.

Meanwhile, what's happening with the supernovae? Currently, the most prolific observing team is an international group called the Supernova Legacy Survey, and their preliminary results on the nature of dark energy are showing a decided preference for... wait for it... yes, the cosmological constant. The group reports that the negative pressure of dark energy seems to have changed by less than 20 per cent since the Universe was about half its current size, a finding consistent with dark energy being the most conservative of the current models – Einstein's Λ. If this is confirmed, it will be a truly remarkable finding. It means that even when Einstein thought he was blundering, he was right. What a guy.

What will happen eventually to the Universe?

If dark energy is a permanent feature of the Universe, we're destined for a long, cold finale. And even if it eventually switches off, there's unlikely to be enough matter in the Universe for gravity to brake the expansion and eventually cause a cosmic collapse, or 'big crunch'.

As the Universe continues to expand, eventually it will run out of hydrogen fuel, and star formation will cease. All the stars will end up as either black holes or exotic, super-dense cinders like black dwarfs, and the contents of the Universe will cool and die. Eventually, on timescales measured in tens or hundreds of billions of years, even the black holes will evaporate. So, there you go. It's something to look forward to.

There is a slightly more serious side to trying to predict what the Universe might look like in the very distant future. It was highlighted recently in some work by Lawrence Krauss and Richard Scherrer in the United States. They looked forward a few thousand billion years, and discovered that the effects of dark energy would be such that most of the Universe would be moving away from us so fast that we would no longer be able to see it. The CMBR, for example, would have disappeared, as would most of the galaxies. Only the members of our own Local Group (see Chapter 9) would be sufficiently tightly bound by gravity to remain visible. The Local Group would have become one of a host of 'island Universes' whose inhabitants would be unable to detect any of the others.

Imagine what cosmologists in the year three trillion would make of this scenario. They would have no evidence for the Big Bang. They would know nothing of the 100 billion or so other galaxies that we can see today. The conclusions they would draw might be quite different from our current picture. The lesson in this is that even from our vantage point, there could have been events in the Universe's past that are undetectable today. We should always be on our guard against such possibilities.

A WALK ON THE WILD SIDE: COSMOLOGICAL EXOTICA

Are there other universes?

We normally define the Universe as being everything we can detect or know about. It includes all of spacetime – whether it's visible to us or not – and all fundamental particles and interactions (or forces). So, can we possibly talk about other universes? Some cosmologists do, and the former Astronomer Royal (and now President of the Royal Society), Sir Martin Rees, coined the idea of 'multiverses'. Other cosmologists prefer the possibility that a single Universe might be so big as to have many facets – of which we see only one. Either way, Rees' eloquent statement that 'what we have traditionally called the Universe may be just an infinitesimal part of reality' is equally valid.

At the heart of the idea of multiple universes is the concept that reality might consist of more than just the three dimensions of space and one of time that we perceive around us. String theory, for example (see 'What was there before the Big Bang?'), postulates that vibrating strings – which look to us like fundamental particles and interactions – actually exist in more than just the four dimensions we can see. The most basic String theory requires the existence of 10 dimensions, and more refined versions have anything from 11 to 26.

All these additional dimensions are supposed to be rolled up, or 'compactified' in such a way that we can't see them – so don't bother going off to look. They may, however, be revealed in collisions of subatomic particles at the highest energies. This is one of the reasons behind the construction of a gigantic machine called the Large Hadron Collider, currently being built at the European nuclear research centre, CERN, on the French-Swiss border. This particle accelerator is expected to start giving results towards the end of 2008.

Rather easier to get your head around – at least as far as multiple universes are concerned – is the idea of 'large' hidden dimensions of comparable size to spacetime, in something called M-theory. No one seems to know what the 'M' stands for, but 'membrane','matrix','master', 'magic' and 'mother' have all been suggested. I think 'missing' might be better, as the theory is more like an empty framework that has just a few panels filled in here and there, where theoretical physicists have glimpsed some aspect of it.

M-theory is an offshoot of String theory that envisages the existence of structures called Dirichlet membranes, D-branes or, simply, branes. The term celebrates a nineteenth-century mathematician with the unenviable name of Johann Peter Gustav Lejeune Dirichlet, who worked on problems to do with the boundaries of objects. So – what's a brane? Imagine a sheet of paper held up in front of you. That's a two-dimensional object situated in a three-dimensional space. M-theory imagines our Universe collapsed onto a three-dimensional brane that occupies a higher-dimensional space called 'the bulk'. In such a theory, it is possible to imagine the bulk being populated by other branes containing other universes.

In an intriguing twist, M-theory suggests that gravity can actually be transmitted through the bulk. Gravity is by far the weakest of the four fundamental interactions of nature (the others being electromagnetism and the strong and weak nuclear forces), and this might explain why. The idea is that it leaks through from one brane to another. A consequence of this would be that one day we might be able to detect gravitational waves coming from another universe.

Other theoretical scenarios envisage vast numbers of universes. String theory alone can give something like 10^{100} different, equally valid solutions – so does each of them have a universe? Some scientists also note that the laws of physics in our own Universe seem extraordinarily well tuned for the existence of life. Tweak the fundamental constants slightly one way and stars never form; tweak them the other and the Universe collapses into itself before anything useful has had time to happen. These ideas have led to something called the Anthropic Principle, which says that the Universe is like it is because we are here to observe it. The inference is that there could be gazillions of other, lifeless universes co-existing with ours that simply don't have the right conditions for stars and planets to form – and for life to evolve.

Where did the energy of the Big Bang come from?

Yes, it's a great question. The standard model doesn't really have an answer for that – we just know that there was an unbelievably large release of energy in the Big Bang. Conventional physics doesn't allow us even to probe the immediate aftermath of the event, since the temperature and density of the fireball far exceeded what present theories can cope with.

Some of the more exotic cosmologies can hazard a guess, however. For example, imagine two parallel branes approaching one another in the bulk (see the previous answer). One of them is ours. What happens if they collide? This scenario was envisaged in 2001 by physicists Justin Khoury, Burt Ovrut, Paul

Steinhardt and Neil Turok, and is called the Ekpyrotic Model of the Universe – from a Greek word meaning 'conflagration'. Nice choice.

The idea is that the energy of the collision is transferred to a Big Bang event in the brane, producing the various phenomena we observe. It's a highly speculative concept, but at least it offers a suggestion of where the energy of the Big Bang came from. Just a good old shunt. Can it be tested? It's possible that the *Planck* spacecraft will detect signatures in the CMBR that would rule out the ekpyrotic scenario (see the introduction to this chapter), but until then, it's still on the cards.

Why bother to do cosmology – isn't it just guesswork?

I hope what you've read in this chapter will convince you that no, cosmology isn't guesswork. We take the best observations we can, combine them with the best theories and accept the explanation that best accounts for what we see. In doing that, however, we make assumptions such as the laws of physics being the same here and now as they were at the beginning of time – and at the other side of the Universe. We do this because, at present, it's our only option – but it also gives us results that are reassuringly self-consistent.

Occasionally, however, a theory comes along that makes you wonder whether guesswork plays a part after all. One such is the No-boundary theory of Stephen Hawking and Thomas Hertog. This takes the intrinsic uncertainties of quantum theory – where probability plays a crucial role – and applies them on the largest scale. It means not only that the Universe has an infinite range of possible histories, but that the way in which we observe it today can dictate which history is correct. Such 'retrocausality' seems completely counterintuitive – if not bizarre. We aren't used to the idea of the present affecting the past. However, it turns out that if you statistically combine all possible histories of the Universe to get the most likely one, it looks a lot like the standard model outlined in the introduction to this chapter.

Like all the best theories, the No-boundary model makes predictions that can be tested by observation. Once again, the forthcoming *Planck* spacecraft observations of the CMBR might do the job. More convincing, however, will be the detection of a possible signature in the spectrum of gravitational waves – when we eventually have the means to observe them. When will that be? Oh – that would just be guesswork...

CHAPTER 11

COSMIC LOOSE ENDS

SOME *REALLY* INTERESTING QUESTIONS

Finally, and inevitably, there are some questions that seem to defy classification. Their common theme is that they have something to do with the Universe, or the way we look at it – or, in some cases, the way Fred Watson looks at it. They range from the banal to the deeply provocative.

Even though you've taken a lengthy trip along the highways and byways of the Cosmos as seen through the eyes and ears of all those thoroughly attentive radio listeners, I hope you'll still find something of interest in these stragglers. They're a mixed bunch, but their common ground is that they have all occurred to an enquiring mind somewhere along the line. To me, that makes them quite special.

So, on that happy note, on behalf of all the people whose questions have made this book possible, I'll bid you farewell. To paraphrase the words of a well-known and much-missed comedy duo, it's goodnight from me – and it's goodnight from them.

UNFINISHED BUSINESS: THE OTHER BIG QUESTIONS

Why should governments spend money on astronomy and space research when there are so many other needy causes?

Back in Chapter 2, we found that there are some good reasons for spending public money on astronomy. We also saw that the amount of government money spent on astronomy is extremely modest compared with spending on other comparable activities. That was brought home to me particularly forcefully a few years ago, when it became clear that the public money spent on staging the 2000 Sydney Olympic Games would have been enough at then-current budgets to run the whole of Australian astronomy for a hundred years.

Astronomers are not used to asking for large sums of money but, as we saw in Chapter 2, there is one particularly big-ticket item on the horizon. It's called the Square Kilometre Array, or SKA, and it's a linked suite of radio telescopes that will probe deeper into the Universe than ever before. This 20-year project will have an eventual price tag of around US$1.7 billion, way beyond the means of any one participating nation. The trick in raising such an awesome amount of money is to have widespread international collaboration, and that is how the SKA project is proceeding. It truly is a global venture.

Telescopes used in ground-based astronomy, whether they're for optical and infrared astronomy or radio astronomy, are usually fairly modestly priced – the SKA is a rare exception.

Space-based astronomy projects are always in a different league, however. The Hubble Space Telescope's successor, for example, will be a 6.5-metre diameter instrument called the James Webb Space Telescope and is due to be launched in June 2013. With an estimated price tag of US$4.5 billion, it will be one of the most expensive projects ever carried out in space-based astronomy. That sum would buy as many as 50 ground-based telescopes with the same mirror diameter. The argument for spending such an enormous sum of money on a single telescope hinges on the unique science it will be able to accomplish from its vantage point in space.

Such expensive space-based astronomy projects are very few in number, and the only reason funding agencies will countenance them at all is that they provide the ultimate demonstration that a nation has arrived in the Space Age – both financially and technologically. Is science the immediate reason for developing prowess in space? Of course not. Defence and commerce provide the real motivation behind the scientific exploits.

In a different league again is the exploration of the Solar System, with both robotic and crewed spacecraft. China, Russia and the United States are among the world powers seeking to colonise the Moon within coming decades, while the United States and Europe have taken the lead in unmanned exploration. Both these big-ticket activities have potentially enormous returns. Imagine the pharmaceutical benefits of a new regime of biodiversity brought about by the discovery of microbial life on Mars, for example. A cure for cancer, perhaps? And the Moon has a million tonnes of the rare isotope helium-3 locked up in its soil, a resource that could meet the world's energy requirements with clean fusion power for the next ten thousand years.

The current US lunar exploration programme is costed at rather more than US$100 billion over some 13 years, a staggering sum. Even though a typical robotic exploration mission costs perhaps 30 times less, both represent colossal amounts of money by almost any standard. On the other side of the balance, the potential benefits for humankind are huge. Projects such as these – human exploration in particular – rank among the largest items of science expenditure of our day. Each is essentially unique, and carries with it the aspirations of all humankind.

So – why should we spend all this money on astronomy and space research when there are so many other needy causes? Even if the reasons suggested above weren't compelling, the bottom line is that it's never a straight choice between one and the other. If it was, one would hope that the immediate humanitarian cause would win hands down – every time. Government expenditure, however, is always a balance of needs, availability of resources and long-term goals, and who knows what needy cause might someday benefit from the exploration of the Solar System, or an understanding of the structure of galaxies?

Is there sound in the Universe?

'In space, no one can hear you scream.' It must be true. Sound needs a medium such as air to propagate, and in space – as everyone knows – there is a vacuum. Well, sort of. In Chapter 3 ('How does the Sun affect communications and power transmissions on Earth?'), we noted that there is indeed stuff in the Earth's environment in space, but that it is extremely thin. This 'interplanetary medium' consists of dust particles, atoms and subatomic particles. Outside the Solar System is an interstellar medium, consisting mostly of rarefied hydrogen and helium gas, small amounts of other gases and minute quantities of dust. The gas comes in quantities of about a million atoms per cubic metre, while the dust amounts to one grain per

100,000 cubic metres. Hardly enough to need cleaning up. But the fact that the particle density of the interstellar medium is something like 30 billion billion times less than that of Earth's atmosphere tells you immediately that there's not much point in listening out for normal sound waves in space.

Surprisingly, however, sound waves – or, more correctly, acoustic oscillations – do have a role in the mechanisms of the Universe. In Chapter 8 ('Do stars shake?'), for example, we discovered that stars such as the Sun oscillate continuously, giving astronomers a unique way of probing their interiors. Very recent investigations have also shown that looping streamers of electrified gas arcing along magnetic field lines hundreds of thousands of kilometres above the Sun's surface (so-called coronal loops) actually guide sound waves like organ pipes. These *basso profundo* musical notes are triggered by explosions on the Sun with the power of a million or so H bombs, known as microflares. An acoustic coupling of the Sun's surface (the photosphere) to its outer atmosphere (the corona) could provide an energy transport mechanism between one and the other, perhaps explaining why the corona is so much hotter (at a million degrees) than the 5,500 °C photosphere.

Another type of sound wave – albeit an unusual one – that plays an important role in shaping the Universe is the density wave that runs through the interstellar medium of spiral galaxies, creating their beautiful spiral arms (see 'Why do galaxies have spiral arms?' in Chapter 9). Technically, this type of wave is known as a soliton, having a single crest that, in this case, triggers energetic star formation in the disc and delineates the spiral arm.

Finally, and most importantly (since without them we wouldn't exist), there are the sound waves that boomed through the infant Universe while it was still glowing brilliantly in the aftermath of the Big Bang. In Chapter 10, we discovered that these waves caused slight temperature variations in the early

Universe that eventually gave rise to the vista of galaxies we see around us today. It does seem remarkable that our own Milky Way Galaxy and, as a result, our own species, owes its origin to nothing more substantial than the echo of the Big Bang.

Is there anything in astrology?

There's no doubt that for most of us, astrology is a lot of fun. I defy even the most hardened sceptic to skip over the astrology page of a magazine without at least glancing at their horoscope. The belief that the positions of the Sun and planets in the sky at the time of a person's birth can influence their lives and careers is an ancient one that probably helped to turn astronomy into an acceptable field of study. One early description of astronomy was 'the handmaiden of astrology'.

At the time of the Renaissance, astrology was still a legitimate scientific pursuit, and any astronomer worth his salt was also expected to be able to cast horoscopes for the rich and famous – or anyone else, for that matter. One of the greatest astronomers of the day – and the last great naked-eye observer of the sky – was the Danish 'Lord of the Stars', Tycho Brahe. In his student days, Tycho practised astrology as a matter of course. As his involvement with astronomy grew, so too did the expectations of others as to his astrological prowess. However, it soon got him into trouble.

An eclipse of the Moon on 28 October 1566 was regarded as a significant event, and the 19-year-old Tycho's considered opinion was that it foretold the death of Suleiman the Magnificent, Sultan of the Ottoman Empire. Since Suleiman was then aged 70, this probably would have seemed like a fairly safe bet. Tycho publicly announced his prediction, but was quickly embarrassed to discover that Suleiman's death had actually taken place several weeks before the eclipse. Ridicule followed – and no doubt continued to follow Tycho around for some time afterwards. It has been suggested that this was the prickly issue that eventually led Tycho into a duel with his cousin, Manderup Parsberg, in which Tycho lost most of his nose. So that's what you get making astrological predictions.

Modern-day practitioners of astrology take great care in their art, making every possible use of today's technology. To an astronomer, however, the fundamental premises on which they base their pseudoscience have no basis in reality. And the statistics bear this out. Time and again, astrological predictions have been demonstrated to have no better success rate than random guesses. That's fine – except when astrologers make claims about their practice that are impossible to substantiate, and use them to extract money from gullible people. Then it's not so funny.

Can we ever know everything?

Probably not. There are good reasons to believe that the pursuit of knowledge will never have an end. One of them is a mathematical theory proved by the logician Kurt Gödel in 1931. Usually known as Gödel's Incompleteness Theorem, it says something to the effect that a mathematical theory can be complete or consistent – but not both. It has echoes of Werner Heisenberg's Uncertainty Principle in physics, which says that at the quantum level (which is exceedingly small), you can either know the position of a particle, or its speed, but not both. Heisenberg's work predated Gödel's by some five years.

Some of today's physicists disagree on the applicability of Gödel's theorem to scientific knowledge in general. In an event at the World Economic Forum a few years ago, two physicists of great eminence, Brian Greene and Freeman Dyson, were invited to debate the question 'When will we know it all?'. Greene argued the case for 'soon', while Dyson took the opposing view: 'never'. Greene, a masterly exponent of String theory, maintained that once we have a 'Theory of Everything', physics will have all the answers. Dyson argued that every new discovery about the Universe reveals a new layer of unanswered questions, citing Gödel's theorem to support his view.

It's worth noting that Stephen Hawking has recently expressed similar sentiments to those of Dyson. And, in a related development, physicists such as Seth Lloyd and Paul Davies have shown that the amount of computational power that could be brought to bear on any particular problem has an ultimate limit – one imposed by the Universe itself. It amounts to some 10^{120} bits of information, the maximum that could be processed in the 13.7-billion-year lifetime of the Universe.

It seems to me that history bears out Dyson's take on this issue, and I'm inclined to agree with him. As a simple example, look at the way our view of time has changed. We saw in the last chapter ('What is space-time?') that Newton's view of time as an absolute quantity meted out in exactly the same way throughout the Universe was overturned by Einstein in his Special Theory of Relativity. That breakthrough allowed us to understand events taking place at speeds close to the speed of light. Relativity tells us that time, like the three dimensions of space, is a dimension and, indeed, the mathematical equations bear this out. But if it's true that time is analogous to space, why can't we move around in time just as we can in the spatial dimensions?

Here's a related question. Does time exist in its entirety? If it does, why can we perceive it only moment by moment? Why does it flow only one way? Other questions come to mind. Is a second of time today the same as a second of time tomorrow? That one was first posed by the mathematician Henri Poincaré more than a hundred years ago, and still has no answer. And, perhaps most provocative of all, some of the exotic theories described in Chapter 10 ('Are there other universes?') suggest that reality might have many other dimensions than the four we can perceive. What's to stop some of those additional dimensions being time-like rather than space-like? The consequences of such a strange mix could be indeed profound.

It's my guess that even when we have a workable Theory of Everything, questions such as these will remain unanswered. They belong to the next stratum of mystery – or perhaps even the one below it. And you can bet your life that once we've got the answers to them (and are happily whizzing backwards and forwards through time), there will still be unanswered questions to come.

Do astronomers believe in God?

To be honest, astronomy doesn't really tell you anything about God. It tells you about the Universe. So, the bottom line is that as with any other group of people, some believe, others don't. I think it's fair to say, however, that astronomers do tend towards the more atheistic end of the spectrum.

There are some aspects of astronomy that clearly challenge certain religious doctrines, particularly the more fundamentalist ones. We know with all the certainty we can muster, for example, that the Universe is 13.7 billion years old, rather than a few thousand years old. This comes from observation and reasoned argument, as opposed to the writings of early mystics who were trying to make sense of the world around them in whatever way they could. Hard-line atheists such as Richard Dawkins might have us throw away all these dusty old books, but that would be to deny humankind an important part of its cultural heritage. While only fanatics would argue that we should accept those ancient teachings lock, stock and barrel, they still rank among the world's literary gems. And what would this world be like without Strasbourg Cathedral, Bach's 'St Matthew Passion', or the Blue Mosque in Istanbul? Religion has inspired humankind to some of its most sublime creations.

One of my close colleagues is a committed Christian and, at the same time, one of the world's leading cosmologists. I was interested to know how he reconciled those worldviews, and intrigued by his answer. He believes there are profound absolute truths, to which our understanding only approximates at any one time. Some of that understanding comes from science, some from the humanities and some from religious faith. He sees his faith as valuable because it takes on board the teachings of a few special people who seem to have had profound insights into fundamental ethical values that might themselves somehow be hard-wired into the fabric of the Universe. On the other hand, he would never push his beliefs on another individual.

This multifaceted view of reality is rather similar to my own, although I would admit to a more atheistic outlook. Perhaps God is somehow represented by the laws of physics – wherever they came from. What is more certain is that the Universe does hold deep mysteries, which science could take decades or even centuries to get to grips with. It's what makes the pursuit of knowledge such a profoundly exciting task, and makes me feel proud to play my small part in it.

DARK SECRETS: FRED WATSON REVEALED

What research are you currently involved in?

Ah. I was hoping you wouldn't ask me this. I've never regarded myself as one of the world's high-flying researchers, for the simple reason that I'm not. But what involvement I've had with research over the years has certainly helped me to keep abreast of the subject and, just as importantly, of its practitioners.

The work I'm engaged in at the moment is broadly similar to what I've been doing over the past quarter century, which is to collect detailed information on very large numbers of objects – whether they be stars or galaxies. The information in question comes from each object's rainbow spectrum, which can be analysed to reveal a barcode of its vital statistics. The technology to do that is something else I've been closely connected with over the years. It involves the use of flexible fibre-optic light pipes and intelligent robots, and it was great to be working in this field when it was in its infancy back in the early 1980s.

In my current research, I'm a RAVEr. That is to say, I'm participating in a project called RAVE, which stands for RAdial Velocity Experiment, and is a 10-nation collaboration designed to measure the speeds and other characteristics of a million stars by 2010. As you can read in Chapter 9 ('How do we know there is dark matter?'), this survey is aimed at some of the most fundamental scientific problems associated with our Milky Way Galaxy. As we pass the 200,000-star mark, it is already promising to revolutionise our understanding of the Galaxy's history. We members of the project will continue to RAVE on, and hope that all those stars don't send us RAVEing mad...

Have you had any 'close encounters'?

After almost half a century of watching the sky, I can't even report a single unaccountable sighting of anything, let alone a close encounter. It's pretty disappointing, really, as a verifiable observation of an alien spacecraft and/or its occupants would be the discovery of the millennium. Even a discarded alien crisp packet would be Nobel Prize material. While most astronomers and space scientists do expect there to be other Earth-like planets capable of sustaining intelligent life in our part of the

Galaxy, very few – if any – would subscribe to the view that we have ever been visited by extraterrestrial intelligence. There is simply no evidence for it.

As we saw in Chapter 8 ('Do you think there is intelligent life out in space?'), telescopic images of Earth-like planets orbiting other stars are currently beyond our capabilities. In a decade or so, however, a new generation of telescopes will allow us to find them and analyse their atmospheres in the search for 'biomarkers' – tracers of life – and, perhaps, intelligent life. That's the closest encounter most of us can expect to have.

Is it true that you used to sing with Billy Connolly?

Back in the late 1960s and early 1970s, I used to perform in the folk clubs of Scotland and northern England, attempting to emulate my heroes in the world of folk-blues with a beaten-up old guitar and a fairly beaten-up voice. Those heroes were musicians such as Davy Graham, Bert Jansch and John Renbourn – all still regarded as legends today – and at one stage I could do a pretty creditable impersonation of all of them.

Then I met up with another guitarist who could actually sing, and we formed a two-person band called The Bradford and East Fife Ready-Mixed Concrete Company. Don't ask me why, but it seemed like a good idea at the time. My erstwhile partner, Kenny Brill, still performs in Scotland, and occasionally sends me one of his exquisite tracks. In return, I send him some of my spaced-out 'Galaxy Redshift Blues' numbers, specially written for 'Science in the Pub' here in Australia.

At the same time as The Bradford and East Fife Ready-Mixed Concrete Company was strutting its stuff, another two-person band was doing the rounds of the folk clubs and we occasionally overlapped at the same venues. They were called The Humblebums, and they seemed to get more gigs than we did. Kenny and I put that down to the shortness of the band's name. The notion that talent might have had something to do with it completely eluded us. The Humblebums were Gerry Rafferty, who had a huge success in 1978 with 'Baker Street' and, yes... Billy Connolly. While Billy's skills as a singer and instrumentalist weren't much better than ours, there was something else that shone through in his performances. And the rest, as they say, is history.

So the answer is no, I never did sing with Billy Connolly. But I did sing *at* him a few times.

AND FINALLY...

Why don't you write a book answering the questions you've been asked on-air...?

Well. What a great idea. I'll get on to it straight away.

FURTHER READING

Aughton, Peter *The Transit of Venus: The Brief, Brilliant Life of Jeremiah Horrocks, Father of British Astronomy* (2004, Weidenfeld & Nicolson)

Blair, David and McNamara, Geoff *Ripples on a Cosmic Sea: The Search for Gravitational Waves* (1997, Allen & Unwin)

Clay, Roger and Dawson, Bruce *Cosmic Bullets: High Energy Particles in Astrophysics* (1997, Allen & Unwin)

Daniel, Christopher St J. H. *Sundials* (1986, Shire Publications)

Desmond, Michael and Pedretti, Carlo *Leonardo da Vinci: The Codex Leicester – Notebook of a Genius* (2000, Powerhouse Publishing)

Einstein, Albert *Relativity: The Special and the General Theory* (1961, Three Rivers Press)

Frame, Tom and Faulkner, Don *Stromlo: An Australian Observatory* (2003, Allen & Unwin)

Freeman, Ken and McNamara, Geoff *In Search of Dark Matter* (2006, Springer-Praxis)

Gamow, George *Mr Tompkins in Paperback* (1965, Cambridge University Press)

Gascoigne, S. C. B., Proust, K. M. and Robins, M. O. *The Creation of the Anglo-Australian Observatory* (1990, Cambridge University Press)

Greene, Brian *The Elegant Universe: Superstrings, Hidden Dimensions, and the Quest for the Ultimate Theory* (2000, Vintage)

Greenler, Robert *Rainbows, Halos, and Glories* (1980, Cambridge University Press)

Hawking, Stephen, with Mlodinov, Leonard *A Briefer History of Time* (2005, Bantam Press)

Hearnshaw, J. B. *The Measurement of Starlight: Two Centuries of Astronomical Photometry* (1996, Cambridge University Press)

Hershenson, Maurice (ed.) *The Moon Illusion* (1989, Laurence Erlbaum Associates)

Lynch, David K. and Livingston, William *Color and Light in Nature* (1995, Cambridge University Press)

Malin, David *A View of the Universe* (1993, Cambridge University Press)

Malin, David and Murdin, Paul *Colours of the Stars* (1984, Cambridge University Press)

Minnaert, M. *The Nature of Light and Colour in the Open Air* (1940, Bell) (and several subsequent editions)

Mizon, Bob *Light Pollution: Responses and Remedies* (2002, Springer)

Moore, Patrick *On the Moon* (2001, Cassell)

Robinson, Andrew *Einstein: A Hundred Years of Relativity* (2005, ABC Books)

Steel, D. *Rogue Asteroids and Doomsday Comets* (1995, Wiley)

Walker, Christopher (ed.) *Astronomy Before the Telescope* (1996, British Museum)

Watson, Fred *Binoculars, Opera Glasses and Field Glasses* (1995, Shire Publications)

Watson, Fred *Stargazer: The Life and Times of the Telescope* (2004, Allen & Unwin)

Webb, Stephen *Out of This World: Colliding Universes, Branes, Strings and Other Wild Ideas of Modern Physics* (2004, Praxis-Copernicus)

Wheen, Francis *How Mumbo-Jumbo Conquered the World: A Short History of Modern Delusions* (2004, Harper Perennial)

White, Michael *Leonardo: The First Scientist* (2000, Little Brown)

ACKNOWLEDGEMENTS

If you're going to write a book based on listener questions from radio talk shows, the first thing you need is listeners. So it's to them that my first thanks go. Without their questions, the book wouldn't exist. And, as you can probably guess, its title is one of them.

Then there are the presenters with whom I've worked. The 'regulars' have their well-earned moment of glory in Chapter 1, but I'd also like to thank Alison Buchanan, Lindy Burns, Chris Coleman, Bernie Hobbs, Kerri-Anne Kennerley, Georgie Klug, Rachael Kohn, Simon Marnie, Steve Martin, Julie McCrossin, Janice McGilchrist, Ian McNamara, Margaret Throsby, James Valentine and Robyn Williams. It has been a privilege, too, to work with producers such as Susan Atkinson, Michael Badcock, Rohan Barwick, Lyndall Bell, Helen Browne, June Cowle, Neva Poole, Amy Sambrook, Kelley Shepherd, Jen Smith and Tony Twiss.

Special thanks go to Richard Glover for having me on his 'Drive Show' from time to time.

The list of people who have supported me in writing the book starts with my long-suffering family, who truly bore the brunt of it. Living with someone permanently attached to a laptop is a pain, and my heartfelt thanks go to Trish, James and Will for putting up with it – again. Thanks, too, to my daughters Helen and Anna and their families in Scotland.

It's a very great pleasure to acknowledge the support of Matthew Colless, Director of the Anglo-Australian Observatory, and the AAT Board. Without that, there would be no radio segments. And to *all* my colleagues on the Observatory's staff – thank you for your support.

The book owes much to conversations with friends and colleagues over the years, and I'd like to acknowledge Peter Abrahams, the late David Allen, Jeremy Bailey, Brian Boyle, Russell Cannon, Brad Carter, Victor Clube, the late John Dawe, Glenn Dawes, Peter Downes, Ken Freeman, Gerry Gilmore, Peter Gray, Sandra Harrison, Derrick Hartley, Tanya Hill, Rob Hollow, Stephen Hughes, Aniello Iannuzzi, Chris Impey, David Kilkenny, John Lattanzio, Charley Lineweaver, Nick Lomb, Malcolm Longair, John Lucey, David McKinnon, Rob McNaught, Sir Patrick Moore, Paul Murdin, Sam Nejad, Peter Northfield, Quentin Parker, Roy Perry, Bill Reid, Victor Richardson, Marjorie Roberts, John Sarkissian, Brian Schmidt, Helen Sim, the late David Sinden, the late Bobbie Vaile, Ken Wallace, John Watson, Alan Watson, Pete Wheeler, Graeme White, Doug Whittet, Trevor Wilson and Reg Wilson.

Although the idea of the book came first from a listener, it was Maggie Hamilton and Ian Bowring of Allen & Unwin who got it off the ground, and I thank them both. It has also been a great pleasure to work on the project with Emma Cotter, Angela Handley, Clara Finlay, Felicity McAlpine and Jo Rudd.

Special thanks go to Jennifer Barclay for her work on the Summersdale edition.

Finally, inspiration has come from many people in different walks of life. People such as Kenny Brill, the other half of the erstwhile Ready Mixed folk-blues band; Jim Cannon, a retired Scottish engineer with some extraordinary achievements to his credit; Ross Edwards, legendary Australian composer; David Malin, the man who changed the way we look at the Universe; the photographer, Frances Mocnik, whose visual perception of the world is strikingly different; and Marnie Ogg of Thrive Australia, whose 'can-do' attitude takes my breath away. Thank you all.

INDEX

accretion 104, 157, 224
Adams, Douglas 178
aircraft speed, and time 54
Aldebaran 187, 189
alien beings *see* life, extraterrestrial
Allan Hills Meteorite 173
Allen, David 8, 10
almanacs 21
Alvarez, Luis 174
amino acids 197
Andromeda Galaxy 208
Anglo-Australian Observatory 8, 9, 10, 203
  2dF Galaxy Redshift Survey 214–215
Anglo-Australian Planet Search 202
Anglo-Australian Telescope 7, 31, 38, 106, 193
  photograph of star trails 48
annular eclipses 140

Antarctica 30
Antennae, The 222
Anthropic Principle 254
Apollo 90–91, 99, 119, 142
arcseconds 182
ashen light 133
  see also earthshine
ASKAP (Australian Square Kilometre Array Pathfinder) 30
asteroids 34, 86, 145, 146, 150
  collision course with Earth 174–176
  occultations 169
asteroseismology 193
astrobiology 139, 176, 203
astrochemistry 196
astrolabes 20
astrology 262–263
astronomers, male/female ratio 40–41
astronomy 7, 14–19, 225, 262
  careers in 39–40
  money spent on 258–260
  and religion 265–266
Astronomy Australia 21
astrophysics 152, 153
Atlantis 107
atmosphere 68, 70–71, 79–80
atmospheric refraction 58, 128
atmospheric turbulence 28–29, 30, 80
atomic clocks 54
aurorae 66, 72, 172
Australian Academy of Science,
National Committee for Astronomy 40
Australian National University (ANU) 231
Australian Sky & Telescope 22
Australian Sky Guide 21

Australian Square Kilometre Array Pathfinder 30

barycentre 191
Bean, Alan 134
Bedding, Tim 193
Bell, Jocelyn 199
Bessel, Friedrich 206
Bethe, Hans 152
Big Bang 156, 211, 217 225, 227–229, 230, 231, 242–243
  before the Big Bang 243
  energy of 254
  location 245
billion, defined 32–33
binary pulsar 199, 242
binary stars 190–192
binoculars 23, 25–26, 72, 103–104
  image stabilisation 26
biomarkers 134, 204, 268
Bishop, Roy 22
black holes 181, 199–201
Bode's Law 158–159
bolides 87
Bondi, Hermann 242
books on astronomy 26
Borra, Ermanno 32
Boyle, Charles 165
Brahe, Tycho 262–263
branes 253
Brill, Kenny 268
brown dwarfs 148, 194–195
Buffy 34, 161
Burbidge, Geoffrey 230

calendar month 121

Cambridge Star Atlas (Tirion) 20
Cannon, Annie 41
Canopus 189
Cassini 93, 164
Catterns, Angela 11
celestial equator 55
celestial poles 185
celestial sphere 35, 132
Čerenkov radiation 234
Ceres 146, 149, 158
Challenger 91
Chandra Space Observatory 93
Chandrasekhar, Subrahmanyan 198
Charon 110, 147
chassignites 172
Chia Chiao Lin 219
Chinese constellations 184
Chinese telescopes 23
Cinzano, Pierantonio 37
circumpolar stars 186
circumzenithal arc 79
Clark, Philip 11
Clarke, Arthur C. 100, 101, 102, 112, 114
  Clarke Orbit 102
clocks 51, 52–53
  near the speed of light 237–238
cloud cover, global 133
Clube, Victor 174
clubs, astronomy 26
clusters
  galaxy 208, 212–213
  globular 196, 201
  star 181
COBE 231

Coles, Peter 232
Colless, Matthew 215
Columbia 91, 107
Coma Berenices cluster 208, 212
Comet Arend-Roland 170
Comet Churyumov-Gerasimenko 175–176
Comet McNaught 169–170
Comet Tempel 1 175–176
Comet Tempel-Tuttle 86
Comet Wild 2 175
comets 34, 72, 85, 86, 150, 169–171, 174
  short-period 161
communications outages 65–66
Concorde 55
conjunctions 167
Connolly, Billy 268–269
constellations 19, 184–185
contact binaries 191
Coonabarabran 38
Copernicus, Nicholas 45
corona, solar 261
coronal loops 261
coronal mass ejections (CMEs) 65–66
cosmic microwave background radiation (CMBR) 227, 228–229, 230–232, 246, 250, 251
  axis of evil 231–232
  ripples 230–231
cosmic rays 233–234, 238
cosmogony 155–156, 158
cosmological constant 248, 249, 250
cosmology 225–232, 241, 255–256
Cosmos (Sagan) 44, 209
Crawford, David 37
Croswell, Ken 26

Crux Australis 185
Cygnus X 200
Czech Republic, protecting night sky 37–38

da Vinci, Leonardo 69–70, 133
dark energy 210, 211, 212, 228, 231, 245, 246–251
dark matter 181, 210, 211, 212–217, 222, 228
Davies, Paul 264
Dawkins, Richard 263
Deep Impact 169, 175
density wave theory 219
dinosaur extinction 174
direct motion 159
Dirichlet, Johann 253
Dirichlet membranes (branes) 253
dispersion 73, 81
distance scale, extragalactic 221
distances, measuring 35–36, 205–206
Dobson, John 24, 208, 212
Dog Star *see* Sirius
Doppler Effect 192, 193
Doppler wobble 202
Drake Equation 205
Drake, Frank 205
dwarf planets 150
Dyson, Freeman 263–264

Early Bird 101, 102
Earth 43–46, 150, 157
  axis 50–51, 55–56, 57, 61–62
  climate and solar activity 154–155
  diurnal rotation 49
  gravity 59–61
  impacts on 174

magnetic field reversal 63–65
name 67, 139
orbital speed 44–45, 49, 50
radiation belts 108, 113
rotation 44, 46–47, 49, 52–53, 54–55
seasons 51, 55–56
seen as disc 108–109
seen from space 134–135
structure 59–61, 62, 63–64
wobble 61–62
earthquake (2004) 61–62
earthshine 70, 133–134
eclipses 139
lunar 122, 131
solar 120, 121, 140–141, 149
eclipsing binaries 190–191
Eddington, Arthur 151
Edgeworth, Kenneth 145
Edinburgh 57
Edwards, Ross 275
Einstein, Albert 200, 234, 238–241, 247–248, 250
Ekpyrotic Model of the Universe 254–255
electromagnetic spectrum 28
elliptical galaxies 218
Enceladus 176
equation of time 51, 59
equinoxes 55–56, 57–58
and hours of daylight 57–58
Eris 148, 149, 150, 162, 163
Eros 175
eta Carinae 189, 190
Europa 176
Explorer 1 90, 92
extrasolar planets 156, 158, 195, 202, 204

Extremely Large Telescopes (ELTs) 31
extremophiles 176

falling stars *see* meteors
fireballs 87
Flamsteed numbers 33
fogbows 76
Fomalont, Ed 242
Fourier deconvolution 203
Friedmann, Alexander 227

galaxies 36, 84, 207, 208
  collisions 222–223
  elliptical 218
  evolution 216, 222–223, 231, 242–243
  number of 209–210
  peculiar velocities 221–222
  rotation curves 213
  spiral arms 218–219
galaxy surveys 215, 250
Galileo 154, 163, 180
Galileo 92
gamma-ray bursts 226
Gamow, George 227, 248
Ganymede 176
Garstang, Roy 37
gas giants 150, 157–158, 160–161, 164
  aurorae on 172
Geminids 85–86
general relativity 240–241, 243–244
General Theory of Relativity 199, 200, 214, 239–240
geomagnetic storms 45, 66
geostationary satellites 100–104, 105, 113
geosynchronous orbit 101

giant molecular clouds (GMCs) 197
gibbous Moon 121–122, 123
Gliese 581 34
Global Oscillation Network Group (GONG) 194
Global Positioning System (GPS) 53
global warming 113
globular clusters 196, 201
gnomons 50
Gödel, Kurt 263
Gödel's Incompleteness Theorem 263
Gold, Thomas 242
Goldilocks zone 142
Gonzalez, Guillermo 141
Gorman, Alice 116
GRACE project 106
Graham, Davy 268
gravitational lensing 214, 216, 227
gravitational waves 199, 241–242, 254, 256
gravity 44, 98, 162–163, 214, 253
  and curved space 238–241
  speed 241–242
gravity assist 100
great circle 55, 132
Great Wall of China 134–135
Greene, Brian 263
Greenwich Mean Time 52, 53

Halley, Edmond 167
Halliburton, Richard 135
halo, 22 degree 77, 78–79
Hawking radiation 200–201
Hawking, Stephen 201, 226, 237, 243, 255, 264
Hayabusa 175
Heavens Above 21, 105

helioseismology 194
Henry Draper Catalogue 33, 41
Herschel, William 33, 147, 154
Hertog, Thomas 255
Hewish, Antony 199
Hipparchus 44, 61, 188
Hitch Hiker's Guide to the Galaxy 178
Hoyle, Sir Fred 227, 242
Hubble, Edwin 220, 228, 248
  Hubble flow 221
  Hubble's Law 220
Hubble Space Telescope 28, 30–31, 93, 209, 216, 259
Hulse, Russell 242
Huygens, Christiaan 163
hydrogen 152, 212, 216

infrared astronomy 8, 42
infrared radiation 28
International Astronomical Union (IAU) 33, 144, 148, 149, 184, 195
International Atomic Time 52, 53
International Dark-Sky Association (IDA) 18
International Geophysical Year 89
International Space Station 11, 91, 92, 106–107, 116, 135
  imaging 107
Iridium 104–105
Itokawa 175

James Webb Space Telescope 93, 259
Jansch, Bert 268
Jansky, Karl 27
Jupiter 71, 81, 82, 148, 150, 242
  Doppler wobble 202–203
  magnetic field 171

moons 92, 150, 176

Keeler, James 164
Kepler, Johannes, laws of planetary motion 36
Khoury, Justin 254
knowledge, pursuit of 263–265, 266
Kopeikin, Sergei 242
Krauss, Lawrence 251
Kuiper Belt 110, 146, 161
Kuiper Belt objects (KBOs) 34, 146, 149, 150, 161, 162
Kuiper, Gerard 145

Large Hadron Collider 253
Large Zenith Telescope 32
leap seconds 52–54
leap years 56
Leavitt, Henrietta 41
Legaut, Thierry 107
Lemaître, Georges 220, 227
Leonids 85, 86
libration 137
life
  evolution of 141–142, 170
  extraterrestrial 176–177, 203–205, 254, 267–268
light
  mass 237
  speed of 179, 232–234, 235–236
light curve 191
light hour/day 224
light pollution 18, 26, 36–39, 185
light year 205–206
lightning 68–69
Lineweaver, Charley 231
Lloyd, Seth 264

Local Group 207, 208, 251
look-back times 219–221
lunar eclipses 122, 131
lunar landings 90–91
lunar laser ranging 137
lunar rainbows 75–76
lunation 121

M-theory 253
magazines, astronomy 22, 25
Magellanic Clouds 207–208
magnetic fields 63–65, 171–172
magnetosphere 65, 66
magnitude scale (stars) 187–189
Malin, David 8, 10, 13, 26
  star trails 46, 47
Mars 81, 82, 86, 150, 157, 158, 172, 176
  meteorites from 172–173
  moons 150
  oppositions 167
  terraforming 115
Marshall Spaceflight Center (NASA) 114
Maunder Minimum 154–155
Maunder, Walter 154
Maxwell, James Clerk 163–164, 232
McNaught, Rob 170
Mercury 81, 100, 136, 150, 157, 158, 161, 172, 238, 240
  conjunctions 166
  greatest elongation 166
  transits 167
mesosphere 69
Messenger 100
Messier's catalogue 34
meteor storms 86

meteorites 83–84, 86–87
  from Mars 172–173
meteoroids 83–84, 86, 87
meteors 69, 82–84
  showers 84–86
  sporadic 84–85
Michelson, Albert 232
microflares 261
Milgrom, Mordehai 216
Milky Way Galaxy 25, 37, 180–183, 190, 196, 207, 208, 215
  history 266–267
  swallowing small galaxies 222
Milligan, Spike 23
Mimosa 185
mini-black holes 200
Mir 91
mirrors, telescope 22, 31–32
molecules 196–197
MOND (modified Newtonian dynamics) 216–217
Moon 20, 21, 73, 86, 118–119, 134–135, 172
  absence of 138–139
  and apparent size of Sun 139–142
  crescent 131–132
  dark side 136–137
  earthshine 133–134
  exploration 259
    see also Apollo
  full 42, 75
  gibbous 122, 123
  isotope resource 259
  length of day 135–136
  moving away 137–138
  orange/red 130–131
  phases 120–124

rings around 76–78
rise/set 124–125
and tides 62–63
transparent 126–127
viewed from different hemispheres 125–126
viewed low in sky 127–128
waxing/waning 124
Moon illusion 127–130
moonlight 42, 190
Moore, Sir Patrick 22, 26, 89
Morley, Edward 232
multiple universes 252–254
muons 238
Murdin, Paul 8

nakhlites 172
naming, of celestial objects 33–34
Napier, Bill 174
Narlikar, Jayant 242
NASA (National Aeronautics and Space Administration) 90, 91, 92, 93, 114, 174
navigation, in space 97–100
neap tides 63
Near-Earth Objects (NEOs) 19, 146, 174
near-field cosmology 226
NEAR-Shoemaker 175
nebulae 25, 34, 153
Orion Nebula 187
Neptune 147, 150, 157, 158, 172
moons 150
neutron stars 197–199, 200
New General Catalogue (Dreyer) 34
New Horizons – A Decadal Plan for Australian Astronomy 40
New Horizons (spacecraft) 93, 109, 110, 163, 180

New Scientist 89, 232
Newton, Sir Isaac 94, 234, 264
  law of gravitation 216–217, 238–239, 240
  laws of motion 216–217
Newtonian telescopes 24
No-boundary theory 255–256
non-gravitational perturbations 99
noon 49
northern/southern lights 66, 72
Norton's Star Atlas 20
nuclear fusion 151–154, 156
nuclear waste disposal 109–110
nutation 62

oblateness 162
Observer's Handbook (Royal Astronomical Society of Canada) 22
Occam's Razor 141
occultations 167–169
Olbers' Paradox 211
Olbers, Wilhelm 211
Oort Cloud 150
Oosterlinck, René 116
oppositions 167, 168
optical astronomy 39, 41
optical telescopes 27, 28–29
  mirrors 22, 31–32
Orana Regional Environmental Plan (REP) 38
Orion 130, 186–187, 188
  Nebula 187
  Sword 187
Orion (spacecraft) 91
Orion Spiral Arm 181
orreries 165

oscillations, acoustic 261
Outer Space Treaty 116
Ovrut, Burt 254
oxygen 151, 152

parallax 206
parhelia 78
Parkes Observatory 30
  radio dish 30
Parsberg, Manderup 263
parsec 205–206
phases, Moon 120–124
photography 9, 17
photometry 118, 119, 191, 194
photons 152, 236–237
Pickering, Edward C. 41
Pioneer 10,11 92, 110, 112, 147, 164
  anomaly 111
Planck 231, 255, 256
planetary motion 36
planetary nebulae 153
planetesimals 157, 158, 162
planets 20, 21, 24
  alignments 165–166
  Bode's Law 158, 159
  defining 147–149
  dwarf 148
  exploration of 164
  extrasolar 34, 156, 158, 195, 201–203, 204
  formation 155–156, 162–163
  orbits 160–162
  order of 145
  rotation 159–160
  software showing positions 21

sphericity 162–163
superior/inferior 166
twinkling 81–82
planispheres 19–21
Pluto 93, 110, 148, 159, 162, 163
demoted 33, 144, 147, 149–150
discovery 147
moons 110, 148, 150
occultation 169
tilt of orbit 161
Poincaré, Henri 264
polar motion 62
Pole Star (Polaris) 47, 61
power outages 65–67
precession 61–62
Principle of Equivalence 239
Prometheus 141
protoplanetary discs 157, 160, 170
protoplanets 157
protostars 157, 194
Proxima Centauri 179–180, 196
Ptolemy 44
pulsars 199
with planets 199
pulsating variable stars 194, 221

quantum theory 244, 245, 247–248, 255, 263
quasars 9, 201, 223–224, 226, 242, 243
quintessence 248–249

radiation, natural 27–28, 134
radio astronomy 41, 171
telescopes 29–30, 197, 205
radio broadcasting 10–12

Rafferty, Gerry 268
rainbows 73–75
  circular 73–74
  lunar 77
  secondary 73
Rutan, Burt 92
RAVE (Radial Velocity Experiment) 267
Rayleigh, Lord 70
red dwarfs 34
red giants 153–154, 179
redshifts 211, 221
Rees, Sir Martin 252
reflecting telescopes 23–24
refracting telescopes 23
relativity 237, 264
  see also General Theory of Relativity; Special Theory of Relativity
religion 141, 265–266
Renbourn, John 76
retrocausality 255
retrograde motion 159–160
Riemann, Bernhard 240
Riemannian manifold 240
ring systems 163–164, 169
rockets 97, 98, 99, 112
Rosetta 175
rotation curves 213
Rubin, Vera 213, 214, 216
Rutherford, Ernest 151

Sagan, Carl 44, 209, 210
Sagittarius Dwarf Galaxy 222, 223
Sagittarius Stream 222
satellite tracking cameras 107

satellites 94–96, 99, 103
communications 101–102, 104–105
geostationary 100–103, 104
multiple 105–106
planetary 135–136, 150
Saturn 81, 93, 150, 157, 158, 172
moons 93, 141, 150, 176
oblateness 162
rings 93
scattering 70
Scherrer, Richard 251
Schmidt Telescope 74
science 17, 89, 259
science communication 8
Science in the Pub 268
scintillation 79–81
Search for Extraterrestrial Intelligence (SETI) 205
seasons 51, 55–56
seeing *see* atmospheric turbulence seismology
Shapley, Harlow 181
shergottites 172
shooting stars *see* meteors
shortest day 58–59
shower meteors 85–86
radiant 85
Shu, Frank 219
sidereal time 49
Siding Spring Observatory 29, 38, 170
light pollution 29
Sigma Octantis 47
simultaneity 234–235
singularities 200, 243
Sirius 71, 81, 187–188
magnitude 189

sky
  at night 72–73
  blueness of 71–72
  transparency of 71
Sky & Space 22
Skylab 91
Skywatch Observatory 37
Sobel, Dava 26
solar cycle 66, 90, 154
solar day 49
solar eclipses 120, 121
  total 140–141
solar flares 66, 154
Solar System 144, 149, 150, 259
  Goldilocks zone 142
  'new' 144
  origins 155–159, 160
  plane of 160–162
  scale 35–36
solar time 49, 53
solar wind 65, 99, 155
solar year 56
solitons 261
solstices 55–56, 58–59
sonic booms 87, 233
sound 260–262
Southern Cross 184–186
space 178–179, 206, 211, 228, 241
  bent by gravity 238–241, 244
  expansion speed 246
  and time 190, 234–236
  within the Universe 243–244
  see also Universe
space debris 115–117, 150–151

space elevator 112–114
space law 115–117
space missions 89–93, 110–112, 259–260
space navigation 97–100
Space Shuttles 90–92, 106
space tourism 91–92, 115–117
space weather 65
spacetime 234–236, 239–240
Special Theory of Relativity 232–233, 236, 237, 246, 264
spectrograph 191–192, 193, 202, 214
spectroscope 164, 212–213
spectroscopic binaries 191–192
speed of light (*c*) 179, 232–234, 236
Spencer, Adam 11
spiral arms, galaxies 218–219, 261
spiral galaxies 183
Spitzer Space Observatory 93
sporadic meteors 84–85
spring tides 62–63
Sputnik 89–90
Square Kilometre Array (SKA) 30, 258–259
star atlases 20–21
star trails 46–48
star velocities 8
star velocity surveys 215, 217, 266–267
star-wheels 19–20, 21
Stardust 169, 175
stars 21–22, 180
  binary 190, 192
  brightness (magnitude) 187–189
  brown dwarfs 147, 194–195
  circumpolar 186–187
  colours 47, 189
  in daylight 71–72

density 196
dispersion of their light 81
distances 35–36, 49, 205–206
formation 155–157
giant 178–179
naming 33–34, 185
number of 209–210
visible with unaided eye 183–184
oscillation 192–194
pulsations 193–194
red dwarfs 34
red giants 153, 179
twinkling 79–80
variable 33–34, 193–194
white dwarfs 153, 189, 197–198
Steady State theory 242–243
Steinhardt, Paul 254
stellar mass black holes 200–201
stratosphere 68
String theory 244, 248, 252, 253
subatomic particles 233, 237–238, 253
summer solstice 55–56, 59
Sun 21, 50, 58, 150, 182
and apparent size of Moon 139–142
effects on Earth 65–66, 154–155
looking at 72
nuclear fusion 151, 154
orbit around galactic centre 182–183
oscillations 192, 193, 261
red/orange 70–71
rings around 76–78
seen low in sky 130
size 179
surface temperature 189, 261

and tides 62–63
sundials 50–52
sundogs 77–78
sunrise/sunset 56–57
and shortest day 58–59
sunspots 154–155
sunspot cycle 66, 90, 154–155
super-massive black holes 201, 223–224
superclusters 208
Supernova Legacy Survey 250
supernovae 34, 198
Type Ia 247
Sydney 38
synchronous rotation 119–120
synodic month 119

Tamworth 38
Taylor, Joe 242
telescopes 22–25, 26, 72, 204
Dobsonian 24–25
magnification 24
mountings 24–25
Newtonian 24–25
optical 27, 28, 29, 31
radio 29–30
reflecting 23–24
refracting 22–23
termination shock 111
Thatcher, Margaret 10
tidal friction 136, 138, 159
tides 45, 136, 137–138
high 62–63
neap 63
time 264

by sundial 50–52
as a dimension 234–236
equation of 51, 58
time dilation 237–238, 241–242
time zone 51, 53–54
Titius, Johann 158
trans-Neptunian objects 146
transits
Mercury 166–167
Venus 36, 166–167
Trapezium 187
troposphere 68
Turok, Neil 254
twilight 72

Uncertainty Principle (Heisenberg) 263
Universal Time 51
Universe 43, 211, 252, 264
before Big Bang 243–244
accelerating 247–248
age 225, 245–246
end of 251–252
evolution of 216–217
expanding 219–221, 228, 229, 230, 242, 245, 246
horizons 244–245
infancy 228, 250
multiverses 252–254
number of stars 209–210
see also space
Uranus 81, 147, 157, 158
axis tilt 173–174
magnetic field 171–172
moons 150
naming 33

rings 169

Vanguard 1 117
variable stars 33
pulsating 193–194
vegetation red edge 135
Venera 92
Veneziano, Gabriele 243
Venus 81, 92, 150, 157, 159, 171
ashen light 133
brightness 71
conjunctions 166
greatest elongation 166–167
near-synchronous rotation 136
tilt of orbit 159
transits of 36, 166
viewing nights 26–27
Viking 92, 172
Virgo Cluster 208
virial theorem 212
visible light 28
Voyager 21, 92, 110–112, 147, 164

Wallace, Patrick 47–48
Watson, Fred 9, 10, 22, 170, 257
and Billy Connolly 268–269
education 39–40
research 266–267
weather 68–69
weightlessness 94, 238, 240
Wheeler, John 200
white dwarfs 153, 189, 197–198
William of Ockham 141
WIMP (weakly interacting massive particle) 216

THE BEGINNER'S GUIDE TO FIXING YOUR PC

How to Solve the Commonest Computer Problems

Peter Neale & Geoff Stevens

ISBN 13: 978-1-84024-596-7

Paperback £5.99

Your computer refuses to work. Before hurling it out of the window or calling the nearest 24-hour techie, consult this book. It could save you hours of frustration, not to mention money and lost work. Arranged in an easy to follow format, this book tackles everything from common problems to major disasters. Whether you are a beginner or a professional, this little guide will ensure you have a solution at your fingertips.

Pete Neale began his career in programming in 1970 as part of an experiment to see if teenagers could understand computers, and he has been glued to his keyboard ever since. He is currently a freelance web programmer.

Geoff Stevens entered the IT industry in 1979 as a COBOL programmer, and in 1983 he moved to IBM as a mainframe administrator. He currently works as a Senior Software Developer.

'Every computer owner should have one by their PC'

David Bugden, Computer Bookshops